Handbook of Sample Preparation for
Electron Microscopy and

Handbook of Sample Preparation for Scanning Electron Microscopy and X-Ray Microanalysis

Patrick Echlin
Cambridge Analytical Microscopy, UK

Patrick Echlin
Cambridge Analytical Microscopy, UK
p.echlin@ntlworld.com

ISBN: 978-0-387-85730-5   e-ISBN: 978-0-387-85731-2
DOI: 10.1007/978-0-387-85731-2

Library of Congress Control Number: 2008942785

© Springer Science+Business Media, LLC 2009
All rights reserved. This work may not be translated or copied in whole or in part without the written permission of the publisher (Springer Science+Business Media, LLC, 233 Spring Street, New York, NY 10013, USA), except for brief excerpts in connection with reviews or scholarly analysis. Use in connection with any form of information storage and retrieval, electronic adaptation, computer software, or by similar or dissimilar methodology now known or hereafter developed is forbidden.
The use in this publication of trade names, trademarks, service marks, and similar terms, even if they are not identified as such, is not to be taken as an expression of opinion as to whether or not they are subject to proprietary rights.
While the advice and information in this book are believed to be true and accurate at the date of going to press, neither the authors nor the editors nor the publisher can accept any legal responsibility for any errors or omissions that may be made. The publisher makes no warranty, express or implied, with respect to the material contained herein.

Printed on acid-free paper

springer.com

For Alexander, Charles, Patrick, Francesca, Maeve, and William in the fond hope that some of them might become scientists.

I am dedicating this book to my six grandchildren and thanking a lot of people for their help and, finally for my wife's patience over the past five years.

Books must follow science, and not science books.

–Francis Bacon, 1657

Acknowledgments

This book would not have been possible without the help of many people. I am privileged to have been working in this field for the past 45 years and am indebted to the many friends and colleagues from all over the world who have listened to my ideas, corrected my errors, and provided practical advice. I am also grateful to the many people and manufacturers who have provided information and illustrations for this book and who are acknowledged in the text.

I am particularly grateful to my colleagues with whom I have taught at the Lehigh Microscopy School for the past 32 years. They have been a constant source of enlightenment and their constructive and critical analysis of my work and writings has been very supportive.

Finally, gratitude to Joe Michael for the use of one of his micrographs as a cover illustration for this book.

Patrick Echlin
Cambridge, January 2009

Contents

Acknowledgments		ix
Chapter 1	Introduction	1
Chapter 2	Sample Collection and Selection	11
Chapter 3	Sample Preparation Tools	19
Chapter 4	Sample Support	31
Chapter 5	Sample Embedding and Mounting	47
Chapter 6	Sample Exposure	65
Chapter 7	Sample Dehydration	97
Chapter 8	Sample Stabilization for Imaging in the SEM	137
Chapter 9	Sample Stabilization to Preserve Chemical Identity	185
Chapter 10	Sample Cleaning	235
Chapter 11	Sample Surface Charge Elimination	247
Chapter 12	Sample Artifacts and Damage	299
Chapter 13	Additional Sources of Information	307
References		317
Index		323

1
Introduction

Scanning electron microscopy (SEM) and x-ray microanalysis can produce magnified images and in situ chemical information from virtually any type of specimen. The two instruments generally operate in a high vacuum and a very dry environment in order to produce the high energy beam of electrons needed for imaging and analysis. With a few notable exceptions, most specimens destined for study in the SEM are poor conductors and composed of beam sensitive light elements containing variable amounts of water.

In the SEM, the *imaging system* depends on the specimen being sufficiently electrically conductive to ensure that the bulk of the incoming electrons go to ground. The formation of the image depends on collecting the different signals that are scattered as a consequence of the high energy beam interacting with the sample.

Backscattered electrons and secondary electrons are generated within the primary beam-sample interactive volume and are the two principal signals used to form images. The backscattered electron coefficient (η) increases with increasing atomic number of the specimen, whereas the secondary electron coefficient (δ) is relatively insensitive to atomic number. This fundamental difference in the two signals can have an important effect on the way samples may need to be prepared. The *analytical system* depends on collecting the x-ray photons that are generated within the sample as a consequence of interaction with the same high energy beam of primary electrons used to produce images.

1. The use of scanning electron microscopy and x-ray microanalysis may be considered under three headings.It is first necessary to understand the actual process of microscopy and analysis. This is not considered here in any detail because my colleagues and I have produced an excellent textbook that covers these processes in great detail (Goldstein et al., 2004).

2. It is necessary to consider how to prepare samples prior to microscopy and analysis. These procedures are the topic of this book.
3. It is necessary to be able to interpret the information obtained from the SEM and attempt to relate the form and structure of the two-dimensional images and the identity, validity, and location of the chemical data back to the three-dimensional sample from which the information was derived. This is a topic of continuing debate.

There are two approaches to dealing with the frequent impasse that may exist between the properties of the sample and the optimal operating conditions of the SEM. We can either modify the instruments so they employ less invasive procedures or we can modify the specimen to make it more robust to the withering beam of high energy electrons. The former approaches are discussed in the book by Goldstein et al. (2004); the latter approach is considered here. With a few exceptions, both approaches are a compromise.

Sample preparation is an absolute prerequisite for microscopy and analysis.

Every specimen that goes into the SEM needs some form of sample preparation. There are no exceptions.

Consider carrying out microscopy and analysis on the components at our breakfast table. After drinking our fresh orange juice, we use a knife to butter our toast before we drink our coffee. The glass containing our juice is composed of a beam resistant, non-conducting, non-crystalline, light element solid. The orange juice and the coffee are wet, non-conducting, biopolymer composed of light element materials that are very bean sensitive. The buttered toast is a thermally and beam sensitive, non-conducting, damp, light element biopolymer. The knife is made of a beam resistant conducting metal that, in spite of being washed, is dirty. The plastic plate is made of a beam sensitive, non-conducting, light element polymer and the coffee cup is a beam resistant, non-conducting inorganic ceramic. All of these objects need some form of preparation before they may be examined properly by scanning electron microscopy and x-ray microanalysis.

The preceding example shows that there is a very wide range of specimen types. For convenience they are divided into six groups on the basis that each group has distinct characteristics, and as a consequence, may require different approaches to sample preparation. This sixfold division is somewhat artificial because many specimens are composed of material from more than one of these groups. For example, a human tooth is composed of hard dry, inorganic material in a biological matrix, a car tire is a mixture of metal and polymers, and deep sea oil is a mixture of brine, mud, and organic material. The six groups are as follows.

1.1 Metals, Alloys, and Metallic Materials

These types of samples occur naturally either as metals or mineral ores and as fabricated components such as nanowires and microwires, cables, sheets, girders, and complex structures such as automobiles and televisions. Metals range in atomic weight from the depleted uranium (Z = 92) component of artillery ordnance, to beryllium (Z = 4) used for specimen stubs. Metals are dry and good conductors of electricity and heat ranging from the high conductivity of silver to the low conductivity of alloys such as Nichrome. They are generally chemically stable and resistant to radiation damage in the microscope column (Figure 1.1).

1.2 Hard, Dry, Inorganic Materials

This important group of samples are generally chemically stable but are all poor electrical conductors. The type of specimens includes ceramics, rocks, geological samples, minerals, particles, microelectronic devices, optoelectronic communication systems, integrated circuits, package devices, and fibers. The group also contains a wide range of manufactured materials such as microelectronic devices, concrete, cement and other building materials, concentrated ores, and glass. Fossil bones and teeth may fall into this category provided they are devoid of significantly important moisture and/or organic material. Living or recently dead bones and teeth are better considered as

Figure 1.1. A steel rail tie from the old Colorado Midland Railroad track (1883–1918) near the Hageman Tunnel in Colorado at 3600m, 1980

Figure 1.2. Serpentine sculpture "Warming myself" by Zachariah Njobo, Zimbabwe, 1999

biological material. Some specimens such as clays, soils, mud, and suspended particles, contain varying amounts of water that must either be removed by drying or converted at sub-zero temperatures to the solid state. This second group of specimens are generally composed of low atomic number materials (Z = 1–20) and are resistant to radiation damage in the microscope column. Provided they are dry, the samples are generally chemically stable but are all poor conductors (Figure 1.2).

1.3 Hard or Firm, Dry Natural Organic Materials

This category consists of polymers of carbohydrates, proteins, and fat derived from living organisms. Close to 99% of the chemical composition of all living material is formed from the low atomic number material (Z = 1–20) and include the major elements (H, C, N, O) and minor elements (Na, Mg, Si, P, S., Cl, K, and Ca). The remaining 1% is made up primarily of Z = 24 to Z = 30 and includes the trace elements Cr, Co, Cu, Fe, Mn, Zn, and I, which are important in various metalloenzymes and coenzymes.

The type of samples allocated to this group include wood and its derived products including timber, furniture, and paper; dried seeds and their manufactured powders, some foods, and plant fibers that form the basis of rope, string, threads, textiles, and fabrics. This category also includes hair, leather, wool, and skin, which are composed of complex polymers of proteins and solid waxes derived from fats. These are all derived materials,

Figure 1.3. Nineteenth century Kilim Prayer Rug. Kapadokya, Turkey, 1976

and although they all are susceptible to varying degrees of hydration, chemical destruction, and biological decay, they do not readily form chemical bonds with small molecules. They are poor conductors and are susceptible to radiation damage. Some specimens are easily wetted; others are very resistant to water (Figure 1.3).

1.4 Synthetic Organic Polymer Materials

The primary feed stock for these synthetic polymers are natural oil, gas, and coal, which provide the source for ethylene, methane, alkenes, and aromatic compounds, and which in turn provide the building blocks for polymers, paints, and plastics. They are hydrocarbons composed of hydrogen and carbon ($Z = 1-8$) and have a wide range of everyday applications in the production of fibers, films, membranes, adhesives, elastomers, and engineering and construction polymers. Manufactured polymers also may exist as composites containing minerals, glass, and metals. They are not readily wetted, and are usually more resistant to hydration and chemical destruction than the previous group of dry natural organic materials. As a group they are very poor conductors and are more susceptible to radiation damage than the previous group of samples (Figure 1.4).

1.5 Biological Organisms and Materials

These are living organisms, and they are composed primarily of light elements ($Z = 1-20$), which also contain a large number

6   1. Introduction

Figure 1.4. Plastic orange juicer

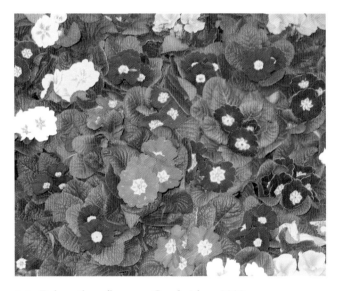

Figure 1.5. Polyanthus flowers. Cambridge, 2008

of trace elements. Water is an essential component of living biological samples, together with complex organic macromolecules such as proteins, lipids, and carbohydrates. They are susceptible to dehydration, chemical destruction, and biological decay. They are poor conductors and very susceptible to radiation damage (Figure 1.5).

1.6 Wet and Liquid Samples

This group of samples, although unique, is a subset of the previous five samples. They are included here as a separate group because their liquids are an essential part of their form, structure and chemistry.

Wet samples are defined as specimens that do not flow and have a readably recognizable structural integrity at normal temperatures and pressures. For example, a drilled rock sample from an oil field, a fresh leaf, or freshly mixed concrete.

Liquid samples will flow and lack any readably recognizable structure at normal temperatures and pressures; for example, paint, mud, and milk (and beer).

In both types of sample the most prominent liquid is water, but anhydrous oils, emulsions, inorganic and organic solutions, and suspensions are found in the natural state of many different specimens. These are the most challenging group of samples to prepare for although there are specialist devices to directly examine wet samples in the SEM, most microscopy and analysis is carried out on dry samples. The liquid components are exclusively composed of light elements and are very susceptible to conventional drying and chemical activity. The liquids are poor conductors and very susceptible to radiation damage (Figure 1.6).

This book is designed as an introduction to sample preparation and will not contain recipes for every type of specimen. The approach is to bridge the gap between the properties of the specimen and the exigencies of the SEM and provide a general

Figure 1.6. Wet yogurt and and beer, 2008

way to prepare the sample. This general approach is continually interspersed with additional more specific information to which readers are directed in the scientific literature.

It is intended to discuss in general terms, twelve different activities that are involved in sample preparation arranged in the sequence they would normally be expected to occur during sample preparation. It will be made clear where there are notable differences to this sequence. For example, metals and dry inorganic samples may need repeated cleaning.

Prior to designing any sample preparation, it is necessary to clearly establish the scientific questions to be answered and whether scanning electron microscopy and x-ray microanalysis are the most appropriate instrumentations to use. At this preliminary stage it is useful to assemble as much chemical and structural information as possible about the specimen available from other sources. As most samples are multiphase, it is probably best either to design the preparative procedures to accommodate the most fragile component or set out different procedures for each of the different components. For example, a plastic insert in a metal rod might need two procedures and the metal and plastic knee replacement in my left leg might need three or even four methods.

The first chapter considers the collection and selection of the sample to ensure it is given a unique identity code that will follow it through the whole procedure of preparation, examination, analysis, and eventual storage. It is necessary to know the dimensions of the specimen stage inside the microscope, as this will determine the size, shape, and form of the specimen to be studied.

The second chapter considers what tools and equipment are needed to prepare the specimen. These can vary from large diamond saws to fine forceps and from freeze dryers to toothpicks. Usually the specimen and proposed investigation dictate the type of instruments that are needed to image and analyze the sample.

The third chapter discusses the different ways samples may be supported in the microscope.

The fourth chapter considers the ways different types of samples may be embedded in, or mounted on, the chosen specimen support.

The fifth chapter provides information about the different ways the region of interest in the sample may be exposed for microscopy and, if necessary, polished for x-ray microanalysis.

The sixth chapter gives details of the different processes that may be used to dehydrate the sample and ensure it is dry when it is placed in the microscope.

The seventh chapter discusses how delicate samples may be stabilized to enable them to be imaged in the microscope.

The eighth chapter considers how any of the six types of samples may be treated to preserve their chemical identity as well as their structure.

The ninth chapter describes the different ways to ensure the sample is clean and free of any materials that will either obscure the surface image and/or compromise its chemical identity.

The tenth chapter provides details of all the methods that may be used to ensure the sample is electrically conductive inside the microscope column.

The eleventh chapter addresses the problems associated with recognizing artifacts and damage that may arise during sample preparation.

The final chapter provides a compendium of additional continuing oral and written information about new developments in sample preparation.

Ideally, once sample preparation is completed the specimen should be immediately placed into the microscope for imaging and analysis. If this is not possible, it should be stored in a clean, dry, secure environment for examination at a later date. All that remains now is to study the specimen—but that is another story.

2

Sample Collection and Selection

Before considering the different aspects of specimen preparation, it is first necessary to understand some general features of the sample and the scanning electron microscope (SEM). This involves collecting and selecting the samples in order to establish their optimal form and dimensions for examination in a particular scanning electron microscope.

2.1 Sample Collection

Usually sample collection is not a problem as investigators bring their samples to the laboratory, but it is necessary to consider the problems involved in collecting and transporting specimens to the microscope. It is important to try to collect a representative sample; for specimens 1 mm^3 or smaller, the best approach is to collect several items. Large specimens may present a problem, and the best that can be done is to take a photograph of the whole sample to show what part of it was collected. For most inert dry samples, such as metals, rocks, geological specimens, wood, paper, fabrics, seeds, and most plastic and polymer specimens, it is only necessary to carefully wrap the specimen in clean dry material and place it into a secure well-labeled container. Powders, soils, and fine metal filings are more conveniently collected in clean containers or plastic bottles. More delicate samples should be wrapped in several layers of fine dry tissue. At this stage, no attempt should be made to clean the specimen other than to remove any excess surface liquid.

Biological samples and wet samples present a number of problems, and the aim should be to collect and maintain them in their natural state. Aquatic organisms should be kept in their natural environment, terrestrial specimens are best maintained in a damp environment, and dry material should be loosely wrapped in soft tissue. Dissected material should be gently covered with surgical gauze. These types of samples should be kept, ventilated or aerated if necessary, in secure well-labeled containers.

Wet and liquid specimens should be maintained in their natural state in sealed containers made of either plastic or glass. Frozen samples, such as ice cream, should be collected and kept under liquid nitrogen or dry ice and brought back to the laboratory. One of the most remarkable low temperature collecting and transport devices was that devised by scientists at British Antarctic Survey to enable them to collect polar ice and snow and transport the intact precipitation back to their laboratory in Cambridge, England.

The key to these collection and transport procedures is to maintain the sample in a condition as close as possible to its natural environment and speedily transport it to the microscope.

2.2 Sample Selection

It is necessary to consider the parameters that govern selecting a suitable sample for examination in the SEM. Size, shape, and proposed sample examination need to be considered.

2.2.1 Internal Dimensions of the SEM Specimen Chamber

The specimen chamber of the SEM is much larger than the specimen region of a transmission electron microscope (TEM). Some SEM stages can accommodate and image over the whole area of an 8-in. wafer and even a hamburger, as shown in Figure 2.1.

Figure 2.1. Picture of a hamburger. Picture courtesy of John Mansfield, University of Michigan, Ann Arbor, MI

There is one spectacular scanning microscope with a chamber big enough to hold a car tire, parts of a jet engine or even a Terracotta Soldier, as shown in Figure 2.2. Further details of this large chamber SEM may be found in the recent paper by Hariharan et al. (2008).

The specimen stages of most scanning electron microscopes are limited to examining every point on a specimen 5–10 cm in diameter.

A number of factors limit the size of the specimen that may be examined in the SEM. The entrance to the microscope chamber must be large enough to allow the specimen to be placed on the specimen stage. For many SEM this does not present a problem because the entire specimen chamber is first vented to atmospheric

Figure 2.2. One of the Chinese Terracotta Soldiers inside the Large Chamber SEM made by VisiTech. Picture courtesy of VisiTech of America

pressure and the stage assembly is exposed in a draw-like fashion. Quite large specimens may then simply be placed onto the stage; the chamber is then closed and evacuated to its working pressure.

In other microscopes, the mechanical parts of the stage assembly are maintained at high vacuum within the chamber volume and the specimen is inserted onto the stage through an air lock. This arrangement limits the size of specimens that may be put into the microscope for imaging and analysis.

It is important to have a good understanding of the internal geometry of the specimen chamber of the SEM being used for microscopy and analysis. The dimension, position, and size of the final lens must be known. It is necessary to know the position of the specimen stage and how much it can be moved in the X, Y, Z, and tilt directions. It is important that the operator knows where the secondary and back scattered electron detectors and the energy dispersive x-ray detectors are located in the chamber and how much these may be moved. The secondary electron detector is usually in a fixed position at the edge of the chamber and the demountable solid state backscattered detector usually surrounds the final pole piece. The position of the backscattered detector may limit the shortest working distance between the sample surface and the final lens to 3–5 mm. The movable x-ray detector is usually located to one side of the chamber and only moved close to the sample during microanalysis. The position and subtending angle of the detector must be known to avoid a large sample crashing onto the front of the detector.

Figure 2.3 shows the relative position of the secondary, backscattered, and x-ray detectors to a flat specimen at three different working distances in a FEI XL-30 SEM. These images were taken using a wide-angle, fixed focus, infrared camera placed at a fixed point inside the microscope chamber. This a very useful lifetime system to ensure that bits and pieces inside the microscope column do not bang into each other. The quality of the recorded images is ideal.

The working distance is measured from the lower surface of the final lens to the top of the specimen. A short working distance is associated with high-resolution images. A mid-working distance is used for most general SEM and, with the FEI XL-30, for x-ray microanalysis. The long working distance is used to provide an increased depth of field in order to view complexly sculptured specimens. The six images in Figure 2.3(A–F) were recorded either using a charge-coupled device (CCD) camera or an infrared camera attached to the inside of the microscope. They show the relative position of the specimen stage to different detectors as a function of working distance.

An alternative and rather unorthodox way of viewing the inside of the chamber is to attempt to record and image an uncoated glass sphere at 15–20 kV. The non-conductive specimen quickly

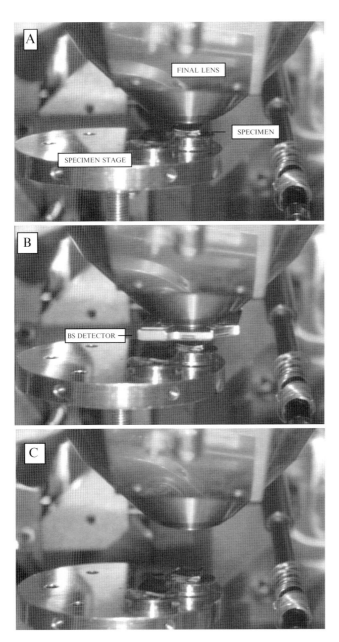

Figure 2.3. A. The settings for a short working distance of 5 mm. The secondary detector in all six figures is out of sight at the top right corner of the images. B. The settings at 7-mm working distance below the back-scattered (BS) detector. C. The settings are at a mid-working distance of 15 mm with no BS detector. D. Settings at 15-mm working distance with the BS detector in position. E. Settings at 15-mm working distance with the BS and the x-ray detector (XR) in position. F. Settings at a long working distance of 30 mm with both the BS and XR detectors in position

Figure 2.3. (continued)

charges and reflects the incoming electrons to the internal surface of the microscope, which provides a highly distorted image that shows the location and relative position of various chamber components.

These types of images, together with the manufacturer's instructions and information, let the operator know how much space is available for a specimen inside the microscope column and how far it may be moved. For example, with a 2- to 5-mm thick flat specimen, the standard stage of an FEI XL-30 microscope can be moved 50 mm in the X and Y direction and the working distance moved 30 to 5 mm. Using a retractable x-ray detector, it is possible to safely examine a 25-mm cube of material.

2.2.2 Microscope Operating Conditions

Having established the spatial parameters of all the components inside the microscope, it is important to establish precisely what sort of information about the specimen is needed. Magnification, resolution, and depth of focus are determined to a large extent by the working distance between the specimen and the final lens and the position of the signal detectors. It is necessary when using a long working distance to examine a large specimen, to avoid touching any internal component inside the microscope. Long working distances reduce resolution but enhance the depth of focus when examining a highly convoluted sample at low magnification. If high-resolution images are required it is necessary to use a much smaller specimen that can be moved closer to the final lens. It may also be necessary to tilt the sample toward the x-ray detector in order to improve the take-off angle of the emitted x-ray photons.

2.2.3 Sample Size

Most specimens examined in the SEM are much smaller than the dimensions mentioned in the previous paragraphs and usually conveniently fit onto the so-called Cambridge specimen stub, which is 12 mm in diameter and 3 mm thick. The general approach is to make the specimen as small as possible without compromising the appearance of the features of interest and the ability of the microscope to image and analyze these features.

2.2.4 Large Specimens

Sometimes it is not possible to cut a very large specimen, such as a piece of rock, down to a convenient size without damaging the material. It can be argued that a large specimen gives a more representative view of the material being examined. Large biological samples have their own set of problems. Such samples invariably need stabilization prior to examination in the SEM. This stabilization is time dependent. The larger the sample, the longer it takes to stabilize. The same problem occurs when it is necessary to dry the specimen. Large porous samples such as some plastics, fabrics, and minerals may take a long time to pump out inside the microscope in order to reach the optimum high vacuum.

2.2.5 Small Specimens

Although smaller samples generally are easier to handle than large specimens, there is one disadvantage to their reduced size. It is important to be certain that the small sample that has been separated from a much larger sample is representative of the whole specimen. This problem is not unique to the SEM but is one of the disadvantages of any reductive analysis process. One way around this conundrum is to examine a larger number of smaller samples taken from the same specimen.

2.2.6 Sample Shape

As far as possible the sample should preserve its natural shape. It may be necessary to examine both the natural surface of the sample and/or details of the interior. The procedures associated with exposing the interior of samples are discussed later, but at this early stage of sample selection it is important to know that both the surface and the interior of the specimen will go into the microscope for examination and analysis.

2.2.7 Sample View

The SEM is designed to obtain information from the very surface of specimens. This information may come either from the natural surface of the specimen or from a surface that has been exposed by artificial means to reveal the interior of the sample. A decision should be made early in the process of sample preparation about the proposed examination of the specimen.

2.3 Sample Labeling

It is vital that the SEM specimen is properly labeled. It is rarely possible to label the specimen directly, but this approach can be achieved with an indelible marker on the underside of a metal specimen. There are a number of marking systems, but an alpha numeric code is straightforward. The sample label should be quite unique and should follow the sample throughout the process of sample selection to the final image or data set. The same unique code should be used in the accompanying mandatory laboratory notebook, which should have recorded details of everything that has happened to the sample.

If it is not immediately possible to label the sample directly, it should be kept in a labeled container. At the stage in the preparation procedure in which the sample is firmly attached to the specimen holder, the unique label should be firmly attached to the underside of the specimen holder. In this way, the identity of the specimen will not be lost, particularly in conditions in which several different samples are placed on the same specimen stage.

3

Sample Preparation Tools

3.1 Introduction

Samples destined for examination and analysis by scanning microscopy come from two different sources. Many specimens are sent from established laboratories that, although fitted with the appropriate equipment designed for the study of all aspects of the particular discipline, may lack the facilities for microscopy and analysis. For example, a materials science laboratory would include a furnace, whereas a biological laboratory would be equipped with an ultra-microtome, although both types of laboratory would contain light microscopes and balances.

The other sources of specimens are those collected directly from the natural environment or from manufacturing facilities and commercial and public centers that lack any equipment for scientific enquiry. Thus, clothing for forensic study, a failed oil pipe, or a newly discovered meteorite require collection and transport to a facility for preliminary treatment before they are prepared for microscopy.

There are two types of SEM specimen preparation laboratories—those that provide a *general* service for any type of sample, and those *dedicated* only for the study of a particular type of specimen. It is inappropriate to provide a long list of all the tools, equipment, and processes used to handle all types of specimens. The prospective samples must first be examined in an appropriate institution to establish the nature of the problem; make a preliminary assessment in terms of size, shape, and form of the likely sample; and most important of all, must determine whether the SEM is the most appropriate way to provide information. Remember that smart scientists only tackle solvable problems.

The more specialized institutions may have some but not all of the tools and equipment for microscopy. Thus, a metallurgical laboratory would have equipment to handle quite large pieces of metal using stretching, compression, bending, fracturing, cutting, and melting devices, together with saws, grinders, and polishing equipment. Some of the same types of equipment would be found in a laboratory handling large inorganic specimens such as rocks, fossils, and smaller samples such as particles, dust, and semiconductors. Hard organic materials such as wood, paper, and fabric rely on cutting, sawing, and fracturing equipment. Softer pliable material, such as many polymers, may only need to be cut to initiate sample preparation. Studies on biological material require a broad range of specialized tools and equipment: Wet and liquid specimens may require using the specialized equipment developed for low temperature microscopy and analysis.

The starting point for considering the essential tools and equipment needed for SEM sample preparation will be with specimens that, by one means or another, are considered appropriate for study in the SEM but that need specialized specimen preparation. The specialized tools and equipment needed to maintain and repair the microscope, x-ray microanalyzer, and image processing and storage devices are not considered.

3.2 Specimen Size

It is important to produce samples that will fit inside a SEM. Chapter 2 shows that some scanning electron microscopes can accommodate very large specimens by using large specimen stages and chambers. Most non-specialized scanning microscopes, although they can handle specimens up to 50 mm diameter and 20 mm thick, usually examine much smaller samples that can fit either on the 12-mm diameter, so-called Cambridge support or stub, or the larger, 25.4-mm (1-in.) stub shown in Figure 3.1. (This and other images are courtesy Frank Platek, US Food and Drug

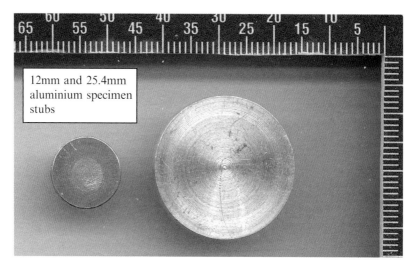

Figure 3.1. 12- and 24.5-mm aluminum specimen stubs

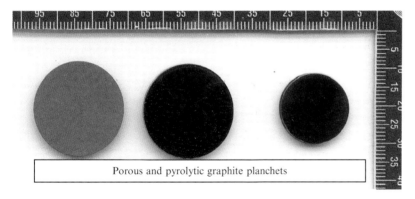

Figure 3.2. 12- and 25.4-mm porous and pyrolytic carbon support planchets used for imaging and analyzing specimens in the SEM

Administration, Forensic Chemistry Center, Cincinnati, OH, although depiction of commercial products in any of the figures supplied by Frank Platek does not imply endorsement by the US Food and Drug Administration. (See Figs. 3.3, 3.5, to 3.8, & 3.11))

The metal support stubs shown in Figure 3.1 are suitable for holding specimens that are to be imaged in the SEM. However, as shown in Chapter 9, more specialized support stubs may be required for specimens that are to be analyzed by x-ray microanalysis. Figure 3.2 shows some of these stubs.

3.3 General SEM Sample Preparation Laboratory

If involved in the construction of a new SEM laboratory, it is useful to first consider the various activities that will take place. It is inappropriate to discuss details of laboratory design and construction here, but the following requirements are important.

There must be separate wet and dry areas; a low-vibration area for sensitive equipment; a region set aside for sputter and evaporative coatings; spark-free storage cupboards and fume chambers; drying ovens and ultrasonic cleaners; plenty of cupboards and shelves; a region set aside for any pieces of specialized equipment; and secure places to tether high-pressure gas bottles, such as those containing nitrogen, carbon dioxide, and argon. The whole laboratory should be well lit and ventilated, and fitted with plenty of electrical power points and appropriate laboratory furniture, including comfortable chairs. There must be separate waste bins for glass and sharps, toxic and hazardous material, and general waste. All these pieces of equipment and the general arrangement and operation of the laboratory must meet the Health and Safety regulations in force in a particular country, as some of the chemicals and many of the operations involved in sample preparation are potentially hazardous.

As far as appropriate, specimen preparation should be carried out on a firm bench with a comfortable adjustable chair, because small hand-held tools must be held securely. For very small specimens it may be necessary to use a micromanipulator to hold and move the tools.

3.4 Equipment to Facilitate Looking at the Process of Sample Preparation

A few larger pieces of equipment facilitate specimen preparation, particularly for smaller samples. It is useful to have a good quality 100-mm diameter glass hand lens with a magnification of ×5 that can be fixed to an adjustable support and is close to a pair of flexible light sources. A smaller ×10 magnification hand-held lens is useful for inspecting the sample more closely. Easy access to a good binocular light microscope is an additional advantage.

Specialist laboratory suppliers provide tools for SEM, but depending on the type of samples being handled, suitable tools also can be found at DIY and hardware stores, opticians, dentists, and model building shops. The following two companies provide specialized tools for use with the SEM. Microtools for Microscopists (www.minitoolinc.com) and Dorn and Hart Microedge Inc. (www.dornandhart.com). Additional information is given in Chapter 13.

3.5 Tools to Expose Samples

3.5.1 Grinders, Saws, and Cutters for Hard Samples

The size of these tools depends very much on the size of the specimen being studied. Choose the size, form, and power for the expected type of specimen. Dental burrs are an excellent way to trim and grind small samples of hard material. Mud, soils, and liquids may contain fine particles of suspended or deposited material, and the filtration equipment shown in Figure 3.3 can be used to separate liquids and solids.

3.5.2 Blades, Knives, and Scissors for Soft Samples

Sharp surgical knives and disposable scalpel blades of various shapes and size are available from a number of different supply houses. Disposable razor blades are very effective for cutting away larger pieces of soft material, but it is important to first wipe the surface with acetone to remove the fine layer of oil. Very thin stainless steel razor blades can be cut to size using scissors and then held in a metal holder.

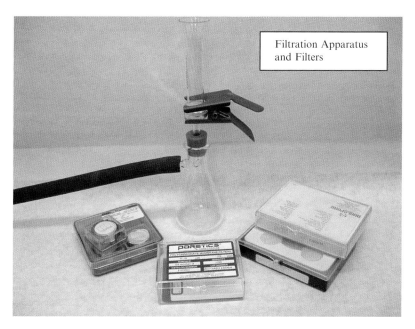

Figure 3.3. Filtration apparatus and filters composed of a system that allows a liquid suspension to pass through filters of various size designed to retain solids of different sizes and compositions

Tungsten carbide knives are harder and tougher and can be used to cut soft metals. Kyocera ceramic knives (www.kyocera.co.uk) have blades made of zirconium oxide, which is second in hardness to diamond. These blades are very sharp and retain their edge for 6 months. Unlike metal knives they are rather brittle and may break when dropped onto a hard surface. The diamond and glass knives used for microtomy generally are not suitable for cutting and trimming samples for SEM, although they are the best way to cut sections of soft specimens. When using thin blades and knives, hold the cutting edge at an angle and as you cut, slowly and continually draw the blade down and toward you to minimize compressing the sample.

Surgical scissors of different size and shape are available; curved blades are particularly useful. One disadvantage of even very sharp scissors is that the blades compress the edges of the sample during the cutting process. If the cutting tools have to be heat sterilized before or after use, it is important to choose steel blades that remain sharp after such treatment. A wide range of microsurgical tools are available that can be used for both cutting and probing very small specimens. It is useful to have spring-loaded cutting edges that automatically retract after the incision is made.

3.6 Tools to Manipulate Samples

3.6.1 Fine-Pointed Probes, Needles, and Brushes

Dental probes are suitable to expose hidden parts of the sample. Figure 3.4 shows that they are not necessarily expensive.

Sewing needles may be used to pick up small specimens and move them to the sample holder. Fractured tungsten wire provides very fine-pointed needles and the fine fresh needles used in acupuncture are sterile and have very fine points. Small disposable needles can be held conveniently in metal holders.

In addition to the metal needles and probes, wooden applicator sticks, toothpicks, fine glass needles, and even some plant prickles, thorns, and spines can be used to manipulate small samples. Figure 3.5 shows a prepared set of disposable cactus needles that may be used to manipulate specimens.

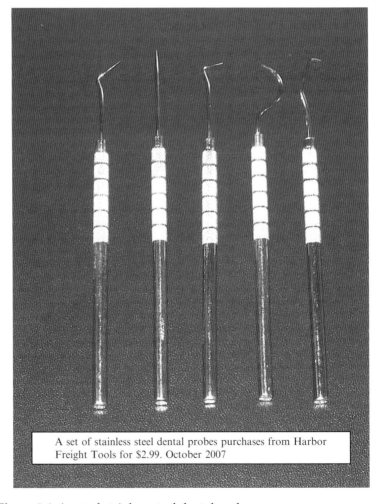

Figure 3.4. A set of stainless steel dental probes

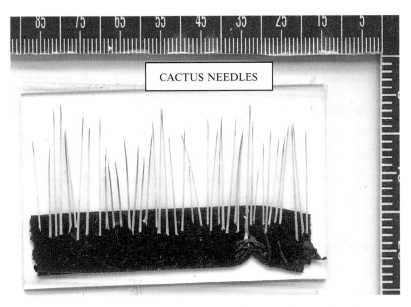

Figure 3.5. A collection of cactus needles that may be used as inexpensive disposable probes

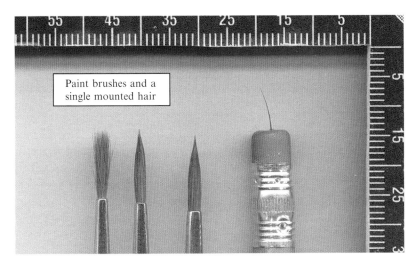

Figure 3.6. A collection of sable hair brushes (bulk and trimmed) and a single pig's eyelash "probe" inserted in a pencil eraser

One-millimeter-diameter soft paint brushes of sable or camel hair and single hairs are very useful for transferring fragments of material from dry specimens. They may be easily cleaned and re-used. Figure 3.6 shows a collection of such brushes.

3.7 Tools to Hold Large Samples

Larger hard specimens may be clamped in a mini vice; soft specimens may be impaled on fine metal points. Forceps are a convenient way to temporarily hold and move specimens. Small and large spring-loaded surgical forceps are capable of holding large samples and picking up and holding small samples. Fine-pointed spring loaded forceps are well suited for picking up particles as small as 100 μm.

3.8 Tools to Help Clean Samples

- *Brushes*. Small 2- to 4-mm diameter natural hair or nylon brushes can be used to sweep away dust and particles on the surface of specimens but they should not be used to paint the sides of non-conducting samples with silver or carbon paint. Wooden tooth picks are much better adapted to this activity. In addition to brushes, it is helpful to have a small container of compressed gas with a fine nozzle that can blow away fine dry particles.
- *Pipettes*. It is sometimes necessary to flush or irrigate a region of the specimen; and fine plastic pipettes with an integrated teat are ideal for this purpose. Fine glass pipettes can be drawn to size from flame-heated glass tubes, but care must be taken not to damage the surface of delicate samples.

3.9 Additional Needs Associated with Sample Preparation

Once a given sample has been selected it is usually necessary to attach the sample to the specimen stub. Chapter 4 discusses the ways samples are supported in the SEM, and Chapter 11 considers the many ways non-conductive samples may be treated prior to imaging and analysis. However, at this early stage it is useful to briefly consider two important components: first to attach specimens to the specimen stub and, second, to ensure there is a conductive pathway from the sample surface to ground. Figures 3.7 and 3.8 show types of conductive adhesive and conductive paint that should be available during sample preparation.

3.10 A Personal Set of Specimen Preparation Tools and Associated Perquisites

Experience has shown that it is convenient if each person working in a specimen preparation laboratory have their own set of general tools in a secure holder. The actual tools in the set depend on the type of specimens most frequently processed, but include scissors, scalpels, saws, forceps, probes, and brushes, and their size is dictated by the usual size of the specimens. The tools for specimen preparation come from a wide range of sources. Figures 3.9 and 3.10 show the author's set of tools and other associated needs for working in the broadly based life sciences

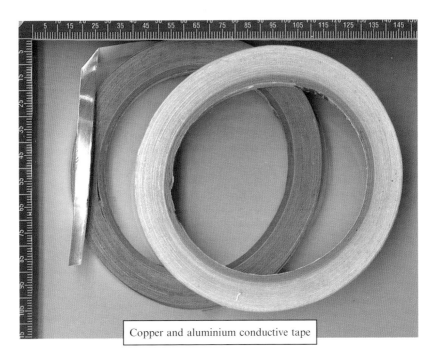

Figure 3.7. Aluminum and copper tape, double-sided and coated with organic adhesive

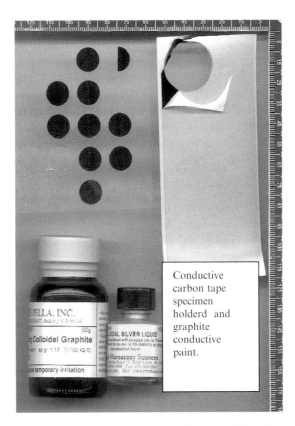

Figure 3.8. Ten-millimeter-diameter carbon mounts with a double-sided coating of an organic adhesive. These adhesive mounts may be peeled off the backing paper and stuck to the surface of a 12-mm stub and the upper layer of paper, and then peeled off to reveal a sticky surface to which a sample may be attached. Two bottles of conductive silver and carbon paint are included in this image. The use of these paints is discussed elsewhere in this volume

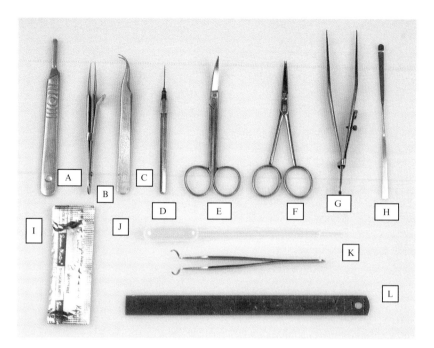

Figure 3.9. A. Scalpel holder. B. Fine forceps with a lock. C. Fine curved forceps. D. Mounted needle. E. Fine curved scissors. F. Fine straight scissors. G. Coarse forceps with a lock. H. Fine scapular. I. Scalpel blades. J. Fine plastic pipette. K. Double curved forceps. L. Ruler

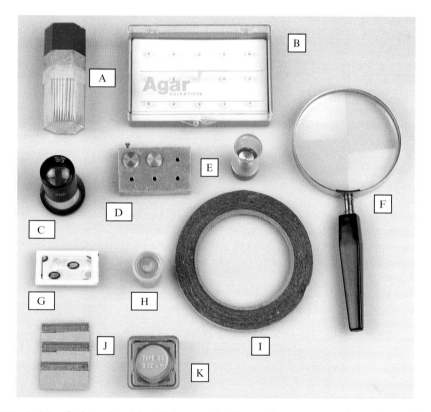

Figure 3.10. A. Wooded tooth picks. B. Large plastic specimen storage box. C. small ×10 hand lens. D. Working aluminum holder for 12-mm stubs. E. Secure container for a single stub. F. Large ×4 hand lens. G. Very thin razor blades. H. Container of Silver-Dag suspension. I. Double-sided conductive adhesive tape. J. Thick razor blades. K. Fine-pore Millipore filters

Multi-Imaging Centre at the University of Cambridge (Figs. 3.9 and 3.10).

3.11 Sample Storage

Once specimen preparation is completed it may be necessary to store the samples either briefly before they are coated to ensure they are conducted, or after coating and prior to microscopy and analysis. There are a number of different storage devices. Figure 3.11 shows some simple secure plastic containers suitable for the temporary storage of 12-mm pointed metal stubs and the flat 12-mm carbon planchets.

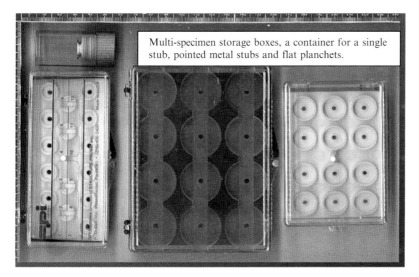

Figure 3.11. *Top left.* A small plastic sealed container is for an individual 12-mm Cambridge stub. *Lower left and right.* These flat storage containers are suitable for storing many 12-mm stubs. *Middle.* The storage box can be used to store flat 12-mm-diameter carbon planchets or flat metal stubs. Specimen boxes have secure lids to ensure sample free of dust and moisture. The boxes must be labeled

4
Sample Support

All specimens need to be supported if they are to be examined and analyzed in the SEM. The support serves a number of different functions:

1. It provides *stability*, so the sample remains in place inside the microscope.
2. It provides a *secure* environment for the sample during the various stages of specimen preparation.
3. It provides a suitable *shape* and *form,* so the sample may be moved through the different phases of specimen preparation.
4. It provides the best *orientation* of the sample during examination and analysis inside the microscope.
5. It enables the sample to be moved out of the microscope and into a secure *storage* device.

It is important to make a decision about the nature of the specimen support early on during sample preparation. For example, cell cultures may be grown conveniently on secondary glass supports that are then stabilized, dried, and processed for charge elimination before being placed in the microscope. They may be attached directly on to a primary support, which may be placed directly into the microscope. In contrast, in order to *image* a metal fracture, it first needs to be thoroughly cleaned and then simply firmly attached to the microscope stage to ensure good conductive contact.

It is convenient to consider specimen supports under three headings:

1. Where an electrically conductive specimen is *self-supporting*
2. Where the specimen is placed on a *primary metal support*
3. Where the specimen is placed on a *secondary non-metallic support,* which in turn is attached to the primary metal support

4.1 Self-Supporting

Some metallic samples may be placed directly on to the metallic microscope stage. This is satisfactory provided the specimen first can be thoroughly cleaned and then firmly attached with a good conductive contact. This seemingly easy approach sometimes is quite difficult to achieve in practice.

4.2 Primary Supports

The primary metal support, usually referred to as a specimen stub, is used to hold the specimen. The composition, form, and shape of the stub is primarily designed to optimize the performance of the microscope. These metallic stubs are also the most convenient supports for samples that do not need very much preparation. Different microscopes have different specimen stubs, although most are either the 12-mm diameter, so-called Cambridge stubs, or the larger 25.4-mm (1-in.) diameter stubs. The stubs can be any size provided they hold the specimen and fit onto the specimen stage.

Most of the 12- and 25.4-mm stubs are machined from either aluminium or Duralium, an alloy of aluminum (90–92%), copper (4%), and minor amounts of magnesium and manganese. The stubs also may be made of pure copper, brass, and stainless steel. Stubs of beryllium or pyrolytic carbon should be used for studies involving x-ray microanalysis.

The metal stubs have a high electrical conductivity and the sample may be placed directly onto the cleaned polished surface or attached either by mechanical devices or chemical adhesives. Figure 4.1 shows some of the different shapes of metal stubs that may be used to support specimens inside the microscope.

The standard metal stubs may be modified easily for different types of specimens. It only requires a little ingenuity and minor workshop skills to change the stubs for a particular type of specimen. For example, fibers can be easily threaded through a pair of holes drilled in the stub; a series of fine needles on the surface of the stub can be used to impale soft specimens; and a shallow dish can hold a small amount of wet or viscous material that is then dehydrated.

Liquid or viscous material also may be spread as a thin layer over the surface of a clean stub or allowed to form drops on a suitably pretreated stub surface. Small dry materials such as dust, particles, pollen grains, metal chips, and microelectric components tend to move around on polished surfaces and most of the sample drops off before it is imaged. One way around this dilemma is to very carefully sputter coat the sample on the stub surface with a thin layer of conductive metal, such as a gold-palladium alloy (see Chapter 11). The thin layer of conductive material provides an anchoring and conductive surface.

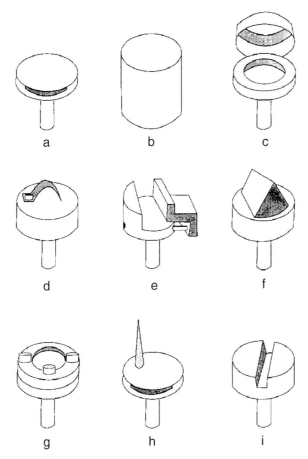

Figure 4.1. Different types of specimen stubs for use in the SEM. Picture courtesy of Judy Murphy (Stockton, CA)

The most effective way to secure small samples is to first coat the stub surface with a thin layer of a suitable adhesive, such as double-sided Scotch Tape. The dry material may be sprinkled on the sticky layer. Slightly larger samples may be viewed under a binocular microscope, and placed at particular locations using fine forceps. This approach is less successful with large specimens because their surface area is much larger than the contact area between the under surface of the sample and the adhesive. Once the material is placed on the sticky surface, the metal stub should be turned upside down and tapped firmly to remove any material that is not securely attached.

4.3 Secondary Supports

The composition and form of the secondary supports are designed primarily to best suit the particular type of sample to be studied in the microscope. The secondary supports are made of a wide variety of conducting and non-conducting materials.

4.3.1 Thin Metal Foils

Thin metal foils are thicker than the thin metal films used for coating specimens, as described in Chapter 11. Depending on the material, foils can be as thin as 400 nm and as thick as 2 mm. A wide range of materials can be used for specimen supports; the foils can be cut to size easily with a pair of scissors provided they are the correct thickness. For example, both copper and aluminum foils 100–250 mm thick can be easily cut to fit the shape of the particular SEM specimen stub. Copper metal foils can be found in some hardware shops and aluminum foil is found in most supermarkets. Goodfellow Metals (www.goodfellow-metals.co.uk) is a good source of a very wide range of pure metal, alloys, and carbon foils in the United Kingdom, United States, and Europe.

4.3.2 Transmission Electron Microscope Grids

The standard 3.05-mm (0.21-in.) -diameter, thin metal discs (or grids as they are called) used in transmission electron microscopy also can be used to hold specimens for scanning electron microscopy. These grids have a wide variety of shapes and configurations and are made either of conductive metals such as copper, gold, and platinum; of low or non-magnetic metals such nickel, rhodium, and stainless steel; of light metal elements such as beryllium and aluminum and non-metallic materials such as carbon and some polymers. The grids are available from a wide range of suppliers including Ted Pella (www.tedpella.com) in the United States and Agar Scientific (www.agarscientific.com) in Europe.

The 3.05-mm diameter grids are 50–100 mm thick, and the central 2 mm are usually penetrated with holes of a wide range of shapes and mesh sizes, or remain intact. The small grids are available in air tight packets and are easily cleaned if necessary. A very popular type of support grid for imaging small sample in the SEM is the 200 mesh copper grid, which has 200 mesh bars per inch. The TEM grids offer a wide choice of supports for SEM specimens in respect to their chemical composition and mesh size. Large particulate material can be placed on the top of woven nylon or metal supports. Small particle and thin sections of plastic and biological material can be supported on fine mesh grids. Non-magnetic material can be examined using stainless steel or carbon grids and samples to be studied by x-ray microanalysis may be examined by using either beryllium or carbon grids. Figure 4.2A–D shows some of the wide variety of grids which may be used to support specimens for electron microscopy.

If the specimens are large enough, the grids are best used without an additional supporting film but, if necessary, the grids

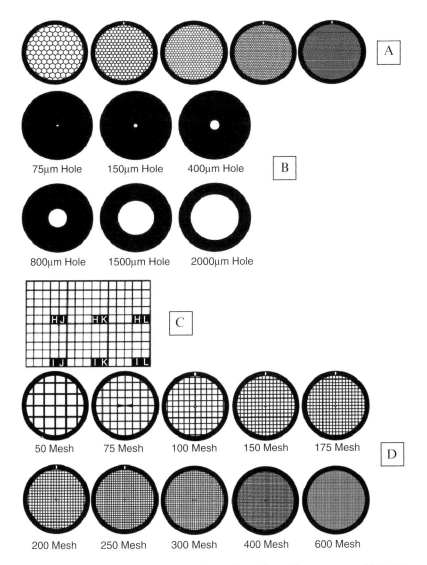

Figure 4.2. A. 3.05-mm hexagonal mesh grids with patterns of 50–400 openings per grid. B. 3.05 holey grids with a range of individual pore sizes. C. 65 mm² locator grid for large light microscope and SEM specimens. D. 3.05-mm square mesh grids with different grid sizes. Images courtesy of Agar Scientific (www.agarscientific.com)

may be coated with a thin support film made of material such as cellulose nitrate (Collodion), polyvinyl-butvar (Butvar), or polyvinyl formal (Formvar).

4.3.3 Thin Supporting Films for Grids

The methods described are an amalgam of different methods described in the literature to prepare very thin supporting films to hold thin sections of material to be examined in the TEM. The thin films for SEM can be a little thicker and, unlike the continuous regular films needed for TEM, also may have a few tears and irregularities.

The general approach is to cast a thin film on the surface of clean water, which is then placed on the surface of the grid. There are three ways this may be carried out:

1. The grids may be positioned on a metal mesh *below* the film surface and the water level slowly lowered until the suspended film is draped over the grids.
2. The grids may be slowly *raised* until they meet the floating film.
3. The grids may be placed directly on the surface of the floating thin film.

4.3.3.1 The Equipment

1. A clean 150 × 80 mm deep Petri dish, fitted with a drain if possible, is needed to float the thin films on the surface. This dish is filled to within 10 mm of the surface with distilled water.
2. If methods (A) and (B) are to be used, construct either a fine wire stainless steel platform about 50 × 80 mm and 15 mm high and place it at the *bottom* of the Petri dish or a similar but smaller device that is held just below the *surface* of the Petri dish. The dish is then filled with sufficient clean, dust-free distilled water until it is a few millimeters below the surface. Ten to twenty of the appropriate type of 3.05-mm grids are placed, shiny side *up*, on the submerged surface of the two types of platforms.

4.3.3.2 Making the Thin Film

1. Thoroughly clean a standard 25 × 75 mm (1 × 3 in.) glass slide in 95% ethanol and dry with a lint-free tissue.
2. A 150 ml solution of 1.0% Formvar in chloroform is placed in a clean glass cylinder approximately 100 mm deep and 40 mm wide.
3. Holding the clean dry glass slide in clean fine-pointed forceps, dip it into the Formvar solution and then take it out and hold it to drain in the vapor phase above the liquid.
4. Remove the coated slide and allow it to dry vertically on a filter paper in a dry, dust-free environment.
5. Place the dried slide on a clean filter paper and using a new razor blade, lightly score around the surface of the slide 3–4 mm from the edge. This breaks the film.

4.3.3.3 Coating the Grids

1. Hold the glass slide bearing the thin film at a narrow angle and slowly lower it into the dish of water and, with luck, the plastic film will detach from the slide and float on the surface.

2. The water level in the container is now either slowly *lowered* or the level of the platform is slowly *raised* and the floating film is deposited on the grids. As the water level decreases it is possible to maneuver the floating film with a clean fine needle so that it lands on the waiting grids. Great care must be taken with this maneuver.
3. If the third method described earlier is being used, the appropriate type and number of grids are carefully deposited, shiny side *down*, directly on to the floating film. The film and grids are adroitly scooped up by pressing a clean glass slide on the floating film and pushing down and upward in a continuous motion.
4. The coated slides should be allowed to dry in a dust-free atmosphere and then carefully removed and placed into a secure container.

A number of laboratory supply companies provide the type of equipment to coat grids with thin plastic support films. Figure 4.3 shows a very simple device which can be made with a tea strainer resting at the top of a 90 × 50 mm deep Petri dish filled with water. A dozen TEM grids are placed, shiny side *up*, at the center of the immersed tea strainer and a 25 mm² plastic film released on the water surface in the same way described earlier. The tea strainer is carefully raised to the water surface to ensure

Figure 4.3. Twelve 3.05-mm 400 mesh copper grids covered with a thin Formvar film and trapped at the base of a 50-mm diameter stainless steel tea strainer

the film layers covers the grids below. The base of the tea strainer is then dried on a piece of paper towel. The cost of this high-tech piece of equipment, which may also be used to make an afternoon beverage, is £6.00 (approx. $9.00).

The single disadvantage to using the standard coated or uncoated grids as secondary supports is that they will not accommodate samples much larger than 3 to 4 mm in diameter. However, even this disadvantage can be offset by the ease by which these supports can be made. With a little dexterity, it is possible to place several grids at specified places on the surface of the primary support. It is prudent to have alphanumeric labels on the primary support to which the grids are held in place by a small amount of a conductive adhesive. Figure 4.2C shows the type of grid that accommodates large specimens. The text books by Bozzola and Russell (1999) and Hayat (2000) contain additional information about the different ways TEM grids may be coated.

4.3.4 Light Microscope Glass Slides

Packets of clean glass slides are readily available in a wide range of shapes and sizes. These should be no larger than 10 mm diameter for 12-mm primary supports and 20 mm for a 25-mm primary support. This allows a small gap at the edge to be coated with a thin layer of conductive material (see Chapter 11). Glass slides are an ideal secondary support for biological samples as it is possible to grow cells and tissue on glass slides and provided the samples are firmly attached to the surface, the glass slides may be passed through the various stages of sample preparation before they are attached to the primary support. The adhesiveness of glass slides can be improved by briefly immersing them in a 0.01% poly-L-lysine hydro-bromide (MW > 300,000) solution. The coated slides should be thoroughly washed in high-purity water, dried, and used immediately. The very thin layer of material imparts a positive charge on the glass surface. Poly-L-arginine and protamine sulfate can also be used in a similar way.

4.3.5 Organic and Metallic Filters and Meshes

Filters are sheets of material with pore sizes within a defined range and are used to separate particles or macromolecules from liquid suspensions. The liquid flows through the pores in cloth, plastic, wire mesh, or pure cellulose paper membranes and the solid insoluble material is retained.

Aqueous suspensions can be retained by using specialist filters made by Whatman (www.whatman.com) and the fused cellulose ester fiber membranes supplied by the Millipore Corp. (www.millipore.com). Thin plastic sheets that have been subject to β-ray bombardment to produce thin uniform channels may also be used as filters.

Fine mesh metal filters are available from the Internet (www.internetmesh.net) and Steriltech (www.sterilitech.net), are used for collecting particles held in suspension in organic liquids that might dissolve some of the organic filters. The mesh pores range in size from 75 to 600 μm and are made of nickel, copper, gold, and stainless steel.

These different types of filters and meshes are available in a wide range of size, shape, pore diameter, and chemical composition. They are a very effective way of collecting organic, inorganic and metal particle, microorganisms, dust, aerosol suspensions, and powders.

To use these filters one needs only to know the pore size of the material to be retained in the filter and to check that the filter and mesh materials are not affected by the suspending liquid. The material collected on the filter is allowed to dry and then cut to fit onto the primary support. The dried filter is attached to the primary support using double sided adhesive tape. This is best carried out by pressed the filter onto the adhesive, not over the whole surface, but at a few point with sufficient force to ensure attachment. This leaves a useful orientation mark on the specimen to show where the surface may be damaged. Figure 4.4 is of an insoluble drug sample collected on an organic filter.

Liquid suspensions may either be dripped onto filters or forced though by pressure. Aerosols and powders are best blown onto the filter surface under low air pressure. The collected particles

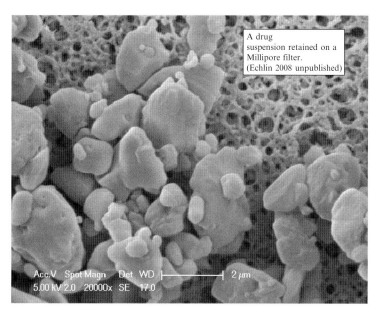

Figure 4.4. An aqueous suspension of a drug sample retained on a 0.22-mm pore size Millipore filter

may be washed, if necessary, before they are carefully dried and cut to fit the surface size of the primary support.

4.3.6 Mineral and Plastic Discs

There is a wide range of these types of material. Freshly cleaved mica can provide a very clean support surface and discs can either be cut to size from thin sheets of a wide range of metals and plastics or purchased direct.

4.3.7 Metal Tubes

One to two millimeter diameter and 2- to 3-mm long metal tubes are slightly tapered at one end and the other end sealed with dental wax. These small tubes, which are best made from high purity copper or silver make very good supports for liquid samples. A small depression is made in the dental wax and is filled using a fine pipette to ensure there are no internal air bubbles and that a 1- to 2-mm droplet protrudes from the surface. The sample may then be quench cooled and is an ideal method for samples that are to be either examined at low temperatures in the SEM or to be frozen dried. The lower tapered end of the tubes is inserted into small holes drilled into a modified primary support. Figures 4.5 and 4.6 show an example of a thin metal support.

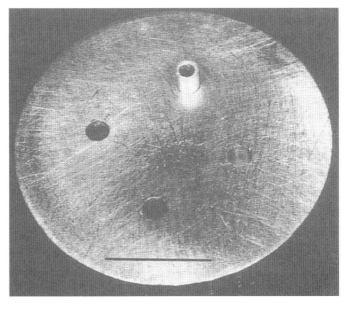

Figure 4.5. A 12-mm specimen stub with small tapered holes for 1 × 3 mm metal tubes. Marker = 5 mm (Echlin, 1992)

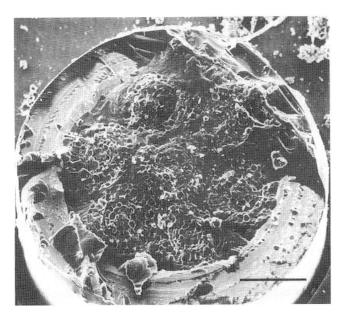

Figure 4.6. A silver tube 0.25-mm internal diameter and 2–3 mm long containing seven quench frozen roots of Duckweed. Marker bar 200 µm (Echlin, 1992)

4.3.8 Fine Hollow Metal Needles

Fine hollow needle probes are an excellent way to take small pieces of biopsy material from living organisms. The device designed by Hohenberg et al. (1996) is shown in Figure 4.7.

4.3.9 Metal Foil Wrappings

Large hard samples such as geological material or precious specimens from museums that have to be returned in their pristine state, can be gently but firmly enclosed in thin metal layer such as aluminum baking foil or copper film. A small space is left free at the top of the specimen to allow the surface of the sample to be examined in the SEM and the little encased sample is secured to the surface of the primary support. Gold foil is probably too thin (and expensive) for this purpose.

4.3.10 Polymerized Plastics

This type of material is an indirect sample support but is a useful way of supporting small particles of some metal and inorganic specimens. The small samples, which are frequently difficult to handle on their own, are embedded in a liquid plastic that is subsequently polymerized. These methods are discussed in more detail in Chapter 5.

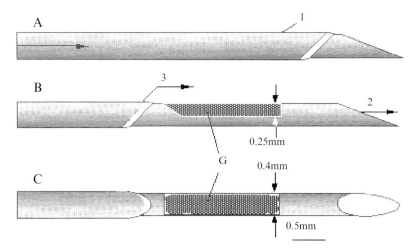

Figure 4.7. Details of a fine needle probe used for taking biopsy samples for electron microscopy. The hollow needle is driven by a piston mechanism into the specimen and the sequence of images (A), (B), and (C) show how the small sample G goes into the hollow needle, from which it may be quickly detached for sample stabilization (Hohenberg et al., 1996)

4.4 Attachments Between Primary and Secondary Supports

4.4.1 Mechanical Attachments

As shown in Figure 4.1, the primary support can be fitted with spring-loaded clips, needle-like projections, or a narrow flat ring of thin metal that fits firmly around the edge of the support. These types of attachments can be used to hold the type of flat metal, plastic, inorganic, and organic secondary discussed earlier. Although these types of holders keep the secondary attachment in place, they do not necessarily ensure that there is complete contact between the two surfaces to ensure there is a good conductive pathway. A firmer and more conductive attachment may be obtained when a circular spherical secondary support is force into a slightly larger tapered hole in the primary support.

4.4.2 Chemical Attachments

There are a large number of paints, glues, and adhesives that can be used to form a firm bond between the sample and the primary and secondary supports and between the two types of supports.

4.4.3 Glues for Dry Samples

A thin layer of a suitable water based or organic based adhesive glue can be easily spread over the surface of the primary or secondary support. The glue may be applied either as a series of droplets or as a thin continuous layer. Not all commercially available glues are suitable for SEM. The glue must not damage the specimen, or contaminate the inside of the microscope. Ideally, the glue should be resistant to the radiation and heat that may be generated by the electron beam. This can be a problem with high resolution imaging of some inorganic samples. The glue must also be easy to apply and form a good smooth tacky surface with an easily recognizable background image when viewed in the microscope. Bozzola and Russell (1999) have made a careful analysis of the many glues that are available and conclude that there are three types of glue which might be suitable for use in the SEM.

Glues based on polyvinyl chloride, alpha-cyanoacrylate, or cellulose nitrate appear to be the most useful. There is a wide range of trade names for these glues in different parts of the world. The layer of glue should be thin enough to hold the sample but not too thick so that the samples sinks and disappears into the wet glue. It is necessary to calculate how long the glue takes to dry to a point at which it is tacky enough to hold the sample firmly.

An effective glue for small samples can be made by simply placing a 150-mm length of 19-mm-wide Scotch Tape in about 5 ml of chloroform. The solvent dissolved the glue on the Scotch Tape and a thin layer can be painted on the surface of a glass or metal support to provide a very effective, quick drying adhesive for small samples.

4.4.4 Paints for Dry Samples

It is possible to make a suspension of fine silver metal dust in either 4-methylpentane-2-one or collodion (a 4% solution of cellulose nitrate in an ether/ethanol mixture). A similar suspension can be made using high purity carbon dust. There are a number of trade names for these paints but they are frequently referred to as Silver-dag and Carbon-dag. These two paints, which are non-viscous liquids, have the advantage of being electrically conductive as well as effective adhesives. They should be applied as a thin layer to the specimen support and allowed to become tacky before the sample is placed on the surface. As is shown later, both silver dag and carbon dag are also used to provide a conducting link across non-conducting parts of the specimen and the underlying metal specimen stub.

4.4.5 Adhesive Tapes

These are a convenient way of attaching samples to supports. The popular brand of double-sided Scotch Tape is composed of a plastic film coated on each side with a thin layer of adhesive. The plastic film can be replaced with a thin layer of a conducting tape composed of carbon, silver, aluminum, or copper. The surfaces of the two sides of the tape are covered with an easily removable paper layer. The tape is cut to size and one of the paper layers is removed and the exposed adhesive layer is pressed on to the support. Once in position, the upper paper layer is peeled off to reveal the second adhesive layer to which the sample may be attached. Although the double-sided adhesive tapes are an easy and effective way to attach specimens to the specimen support, they do have a number of disadvantages.

The two layers of adhesive are non-conducting and will interrupt the pathway of the electron beam from the point of impact on the sample to ground. This potential problem can be easily overcome after the adhesive has dried, by simply painting a thin silver dag pathway from the edge of the upper adhesive layer on which the sample sits, to the metal support to which the lower adhesive layer is attached. This precaution also applies to double-sided adhesive conductive tapes. The conductive pathway should be applied before the whole sample is coated with a thin metal conductive layer.

The tape adhesives are organic materials that can take a long time to out-gas in the microscope column and may cause specimen contamination. The adhesives are also beam sensitive, which give rise to further contamination and may trap gas bubbles between them and the primary support and the stub. The trapped gas expands in the high vacuum and causes instability in the specimen. Although all these problems can never be removed entirely, they can be lessened by some good housekeeping. Once the sample has been attached to the primary or secondary support and a conductive paint layer applied at the edges of the sample, the specimens should be put in an oven at between 30 and 40°C, preferably under vacuum, and allowed to dry out overnight. Only then, if necessary, should the specimen be coated.

4.4.6 Bio-organic Materials

There is a wide range of natural materials that act either as weakly bonding adhesives or strongly binding glues. The plant derived materials are water soluble proteins or carbohydrates and are frequently species specific. Gluten, the main protein of wheat, although practically insoluble in water, can be used as an adhesive. Animal material, such as gelatin derived from skin, connective tissue, and bone, can be treated to form a glue that has strong binding properties. Mention has already been made of the adhesive properties of the polypeptides, poly-lysine, and poly-arginine and the simple protein, protamine sulfate. The distribution and properties of natural glue and adhesives warrant further study as they may well provide new materials for use in the preparation of specimens.

There are a large number of different ways samples may be attached to the specimen supports, but we need to briefly consider the actual mechanism of getting the sample on to the support.

Particles, powders, and dusts can be sprinkled or blown on to the surface and poorly adherent material removed by first taping the surface before blowing it with a jet of dry nitrogen or clean air. This gentle burst of clean gas can also be used to press the specimens further into the adhesive glues.

Large specimen are usually placed at specific places on the support using fine forceps or tweezers and then gently pressed, if possible, to ensure a firm bond. Care has to be taken with attaching secondary supports on to a primary support.

5. Sample Embedding and Mounting

5.1 Introduction

Chapter 4 considered the different ways a sample may be attached to the stage of the microscope. This chapter is an extension to the process of attachment and primarily considers the procedures we need to firmly hold the sample if we want to measure and analyze the inside of a specimen. At first glance, the term embedding, which implies to enclose something closely, is a contradiction of terms for the SEM that aims to image and analyzed exposed specimens. However, an alternate definition for embedding is to make something an integrated part of the whole, and this is precisely what we do when we embed sample for the SEM. There are two forms of embedding, or mounting, as it is sometimes referred to in the material sciences.

5.2 Embedding

The embedding media, usually a liquid, must first thoroughly infiltrate a permeable specimen before it hardens to form a solid matrix. This is what we do with most biological materials that are not sufficiently firm to be cut to reveal the inside. The same embedding principle can be used with small particles of metals, inorganic and organic material, and even porous minerals.

5.3 Mounting

Alternatively, liquid or solid embedding material is used to form a tight bond with the exterior of a largely impermeable specimen before it is polymerized. This enables small and/or thin specimens to be held firmly so they may be cut, fractured, eroded, and polished. This is a common procedure used with metallic and geological samples.

The common practice is to use small samples and carry out the actual embedding/mounting in either small, disposable plastic containers or gelatin capsules, both of which may be sealed. Alternatively, open shaped silicone rubber molds or aluminum dishes may be used. When the emphasis is on mounting large samples, i.e., geological and metallic specimens, larger open plastic or metal dishes may be used. Once the resin polymerization is complete, the specimen may be removed from the container, if appropriate.

The following examples show where embedding and mounting is a useful preparative technique:

1. The outside of a small impermeable geological specimen may be surrounded by a matrix of liquid resin, which is then polymerized to a solid. The solid material may then be cut, ground, and polished to reveal the interior of the sample for subsequent imaging and analysis.
2. Small pieces of a polymer may be embedded in a liquid resin which is subsequently polymerized to a solid. The soft but firm material can be sectioned or planed to reveal features of the embedded polymer.
3. Small pieces of intact metal, metal filings and even powder may be surrounded either by a dry plastic powder or a liquid matrix that is then polymerized, sawn open, ground, and polished to reveal the specimen.
4. Liquid resin embedding and subsequent polymerization is an important prerequisite for examining the interior of soft biological material, either by subsequently sectioning or fracturing the firm plastic matrix.

Embedding for biological samples had its origins in light microscopy, in which warm paraffin wax was infiltrated into stabilized and stained specimens that were then sectioned as thin as 1–2 mm for histological and structural examination. Resins were developed for use in the same way for the TEM to enable sections as thin as 30 nm to be cut from suitably prepared material.

The embedding procedures described in this chapter for biological specimens are derived from extensive investigations for use with the transmission electron microscopy. The books by Hayat (2000) and Glauert and Lewis (1998) contain a wealth of embedding methods, although developed primarily for the TEM, they can be readily adapted for use with the SEM. The manual by Petzow (1978), the books by Van der Voort (1984) and Bousfield (1992), and the publications of the American Society of Metals are full of practical methods for the embedding and mounting metallic specimens. Practical methods for dealing with geological and ceramic specimens can be found in the books by

Buehler (1973) and Lee and Rainforth (1994) and the book edited by Holt and Joy (1989) provides procedures which can be used with microelectronics and semiconductors.

There are a large number of chemicals that may be used as embedding agents, and their suitability for a given sample must meet the following criteria:

1. The embedding material must have a sufficiently low viscosity to enable all exterior parts and, where appropriate, all interior parts of the sample to be infiltrated.
2. The embedding material must form a firm attachment to the sample.
3. The embedding material must preserve the fine structure of the sample.
4. The embedding material must retain the chemical identity of the sample.
5. The polymerized material must be stable in the electron beam.
6. The final embedded material must show no shrinkage and be sufficiently firm to allow post-embedding processes to be carried out.
7. Ideally, the embedding chemicals should be of consistent quality, inexpensive, and readily available.
8. The embedding process should not take a long time to complete.

5.3.1 Embedding and Mounting Hard, Dry, Impermeable Specimens

The methods described here are generally for non-biological specimens. The procedures used may involve heating up to 500–600 K and at pressures of up to 30 MPa. They may also take place at ambient temperatures and atmospheric pressure. Most of the polymerized resins are vacuum compatible in the SEM but if uncertain, first pump out the treated specimens in the vacuum evaporator being used to coat the sample.

5.3.1.1 Hot Methods

These methods use thermosetting materials that flow when heated under high pressure and polymerize to form a stable material. These methods are used to embed small metal samples, metallic fragment, hard dry inorganic samples such as minerals, rocks, archaeological specimens, and semiconductors. All the samples must be unaffected by high pressures and temperatures.

Bakelite, a phenol-formaldehyde resin, is available as fine pellets. The websites www.2spi.com and www.emsdiasum.com have full details of these and other resins and formulations for their application.

The specimen is arranged in a shallow metallic mould with an appropriate amount of resin pellets and placed into a mounting press. It is heated to between 400 and 500 K at pressures of up to 29 MPa (4,200 psi) for 8 min. The thermosetting Bakelite flows under heat and pressure and polymerizes to its final structure. The embedded material, which generally should be allowed to cool at high pressure, is hard and can be readily abraded during subsequent polishing. The resin has good edge retention, i.e., covers all the edges of the embedded sample. This is a quick and inexpensive embedding method for hard, non–temperature and pressure sensitive materials. The Bakelite pellet size and the temperature and pressure of the mounting press should be varied depending on the nature, size, and shape of the sample. The electrical conductivity of the Bakelite resin is poor, but this may be overcome by including a small amount of metallic iron with the resin.

Diallyl phthalates can be used in the same way as Bakelite. It is cured at between 550 and 570 K at a pressure of up to 21 MPa (3,000 psi) for about 5 min. It is a very good embedding and mounting material for metallographic studies as well as for embedding ceramics and glass. It forms a hard resin with low shrinkage and excellent edge retention and can be abraded to a highly polished surface. The resins also may be formulated to contain a small amount of glass granules to increase its polymerized hardness and copper powder to enhance its electrical conductivity.

5.3.1.2 Cold Methods

These methods use a liquid resin with varying amounts of hardeners, softeners, and plasticizers together with an accelerator that is mixed with the specimen and allowed to slowly polymerize at temperatures no higher than 300 K. The cold methods are the most usual way biological materials are embedded. Further information about these materials may be found in the two websites given in the preceding and at www.agarscientific.com. The viscosity of the unpolymerized resin at 298 K varies from as high as 3,000 cPa (centipascal) to as low as 8 cPa. In common with nearly all organic polymers, these resins are non-conductive and susceptible to beam damage by electron radiation. Two general types of resin may be used.

Epoxy resins can be formulated to have a low viscosity that will quickly surround even the minute pores of small specimens and form a good attachment. Higher than room temperatures are needed to polymerize the resin and sensitive samples may need to be cooled. It is possible to use low vacuum infiltrations to ensure complete infiltration of very porous samples. The resin usually takes about 8 h at room temperature to completely polymerize and harden, but this process may be accelerated by raising the temperature to 348 K. The speed of polymerization can be controlled by the amount of hardener used in the resin mix, the more the hardener, and the faster the cure. The edge retention and abrasion rate are very good, and the cured resin has very low shrinkage. There are a number of different formulations to enable one to vary resin viscosity, polymerization time, and final hardness. The polymerized resin is a poor electrical conductor, although iron metal dust may be added to the resin to improve its conductivity.

The epoxy resins are generally immiscible with water and may be used to embed hard dry organic materials such as wood, fibers, and fabrics, as well as the hard dry inorganic specimens mentioned in the preceding. They also can be used to embed engineering resins and plastics and some composites, provided the liquid resin does not cause changes to the polymeric material. Many different types of epoxy resins are used to embed biological material primarily for morphological and structural investigations.

Acrylic resins are similar in many respects to the epoxy resins and involve adding either a dry powder or liquid accelerator to the liquid resin. The flow rate of the liquid resin is not as good as the epoxy resins, but the prepared resin hardens within a few minutes and may be ground and polished shortly afterward. There are a number of different acrylic resins. The water impermeable acrylic resins may be used to embed biological samples for structural studies, and the specialized water soluble resins are used in connection with analytical studies of biological material.

5.3.2 Embedding and Mounting Soft, Moist, Permeable Specimens

The methods described here are largely designed for biological specimens. The procedures are carried out at atmospheric pressure and, depending on the resin, at between 203 and 333 K. Most of the polymerized resins are vacuum compatible in the SEM, but if uncertain, first pump out the specimens in the vacuum evaporator being used to coat the sample. The embedding procedures for soft permeable specimens fall into three main groups.

5.3.2.1 Epoxy Resins

Epoxy resins were introduced to electron microscopy 50 years ago in Cambridge by Audrey Glauert and her colleagues (1956). Consult the excellent description of the use of this material given in Chapter 6 of the book by Glauert and Lewis (1998).

The epoxy resins are generally immiscible with water and can only be used with naturally dry specimens or samples that have been dehydrated. The viscosity of the un-polymerized resin mixture is variable and, like most polymerized organic materials, they are unstable in the electron beam. There are a number of different epoxides and their chemistry is well understood. The epoxy resins used in embedding specimens for electron microscopy are a synthetic thermosetting resin containing epoxy groups and three additional components; a hardener, a plasticizer, and an accelerator.

There are many different types of epoxy resins, hardeners, plasticizers, and accelerators that may be put together to form different epoxy resins; all with different names. These variations are a consequence of the chemical nature of the components, their chemical manufacturer, the viscosity of the resin mixture, and the desired hardness of the final polymerized resin block. To make things even more complicated, some of the original resins and associated chemicals are either no longer available or removed because they are too toxic and carcinogenic. It is best first to contact the various chemical supply houses to find out which epoxy resins are available for embedding for electron microscopy. The epoxy resins are hydrophobic, and specimens must be thoroughly dehydrated before they are embedded. Chapter 7 provides a number of different dehydration procedures, and the Table 5.1 gives a schedule that works well for most specimens.

The infiltration should be carried out in a securely sealed container in a fume cupboard and the removal/replacement of the different materials made with a pipette. The sample should never be allowed to dry out between transfers. With large impermeable specimens, the embedding container may be gently shaken (but not stirred!). During the final stages of embedding, the top of the container should be removed to allow the last traces of propylene oxide to evaporate.

Araldite is an aromatic, low-toxicity epoxy resin originally introduced by Glauert et al. (1956). The viscosity at 298 K of the un-polymerized resin mixture may be as high as 2,500 cP (centipoise) or Pa · s (Pascal second). The addition of various reactive plasticizers to the resin mix can reduce the viscosity. A typical formulation for Araldite is shown in Table 5.2 for Araldite GY 502 embedding medium that contains the following chemicals and is made up as follows (Figs. 5.1–5.5).

The Araldite, DDSA, and DPB are mixed first and may be stored at 277 K in a tightly sealed container for several months.

Table 5.1. Standard infiltration schedule for epoxy resins

Dehydrating agent: propylene oxide	1:1	10 min
Propylene oxide		10 min
Propylene oxide: epoxy resin medium	1:1	1 h or longer
Epoxy resin medium		Overnight
Epoxy resin medium		2 h or longer

From Glauert, 1998.

Table 5.2. Araldite 502 embedding medium

Chemical component	Amount (ml/g)
Araldite 502	19.0/22
DDSA (hardener)	21.0/21.0
DBD (plasticizer)	0.6/0.6
BDMA (accelerator)	1.2/1.2

BDMA = benzyldimethylamine; DBP = dibutylphthalate; DDSA = dodecenylsuccinic anhydride. From Glauert, 1991.

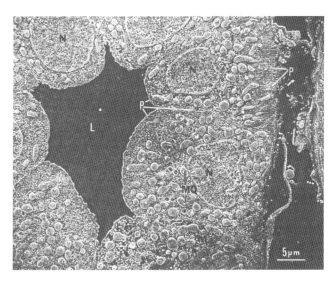

Figure 5.1. An SEM image of a 1-mm thick section of part of a mouse kidney. Stabilized, dehydrated, and embedded in a mixture of an Araldite-Epon resin (Goldstein et al., 1992)

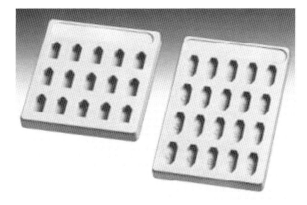

Figure 5.2. Plastic embedding trays. 11 mm long, 6 mm wide and 5 mm deep. Picture courtesy of Agar Scientific (Stansted, UK)

5. Sample Embedding and Mounting

Figure 5.3. Disposable aluminum trays for embedding large samples. Picture courtesy of Agar Scientific (Stansted, UK)

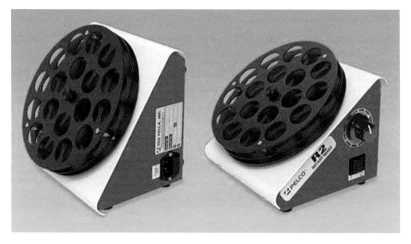

Figure 5.4. Variable speed rotary mixer for use during sample stabilization and embedding. Picture courtesy of Ted Pella (Redding, CA)

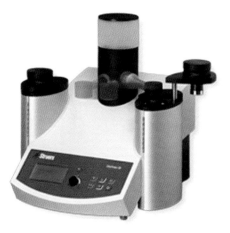

Figure 5.5. A high temperature, high pressure, mounting press suitable for small metallic and hard dry inorganic specimens. Picture courtesy Struers A/S (Ballerup, Denmark)

The DBP acts to reduce the viscosity of the resin. This mixture must be allowed to reach room temperature before it is used. The BDMA is added just before use and is well mixed. The dehydrated sample is passed through the standard infiltration schedule given in the preceding and allowed to polymerize overnight at 333 K.

Epon is an aliphatic, low-toxicity, medium viscosity (550 cPa at 298 K) resin introduced by Luft (1961) as an embedding medium for electron microscopy. It contains the following chemicals and may be made up as follows for soft, medium, and hard mixtures.

It is possible to vary the hardness of the resin by altering the amounts of the NMA and DDSA. More NMA makes a harder resin, more DDSA makes it softer.

The sample may be infiltrated using the standard schedule given in Table 5.3 and the resin is cured overnight at 333 K.

Spurr's resin (Spurr 1969) is based on a cyclo-aliphatic di-epoxide that, although very toxic and carcinogenic, has very low viscosity (60 cPa at 298 K). The advantage of this resin is that it rapidly infiltrates the sample and is particularly useful for specimens made up of hard and soft material. A typical formulation for Spurr's resin embedding medium is given in Table 5.4; it contains the following chemicals and may be made up as follows.

Some of the resin components are anhydrous and readily absorb moisture from the air. These components should be stored in a very dry environment. The resin should be prepared in a fume cupboard just before use, and the appropriate amounts

Table 5.3. Epon 812 embedding medium

Chemical component	Soft (ml/mg)	Medium (ml/mg)	Hard (ml/mg)
Epon 812	20/24	20/24	20/24
DDSA (softer)	22/22	16/16	9/9
NMA (hardener)	5/6	8/10	12/15
BDMA (accelerator)	1.4/1.5	1.3/1.5	1.2/1.4

BDMA = benzyldimethylamine; DDSA = dodecenylsuccinic anhydride; NMA = nadic methyl anhydride. From Luft, 1961.

Table 5.4. Spurr's embedding medium

Chemical component	Firm mixture	Soft mixture	Rapid polymerization
VCHD (resin)	10.0 g	10.0 g	10.0 g
DER (flexibilizer)	6.0 g	7.0 g	6.0 g
NSA (hardener)	26.0 g	26.0 g	26.0 g
DMAE (accelerator)	0.4 g	0.4 g	1.0 g

DER = diglycidyl ether; DMAE = dimethylaminoethanol; NSA = nonenylsuccinic anhydride; VCHD = vinylcyclohexene dioxide. From Spurr, 1969.

Table 5.5. Infiltration schedule for Spurr's resin

Chemical compound		Time
Dehydrating agent:Spurr's medium	1:1	30 min–2 hour
Dehydrating agent:Spurr's medium	1:3	30 min–2 hour
Spurr's medium		4–6 hours
Spurr's medium		Overnight

From Glauert and Lewis, 1998.

(by weight) of VCHD, DER, and NSA placed in a disposable polypropylene container and gently mixed. The appropriate amount of DMAE is added to the mixture and stirred. It is important to cover the embedding molds before they are polymerized to avoid absorption of moisture and loss of volatile components. The resin is infiltrated at room temperature and cured overnight at 333 K. A typical embedding schedule is given in Table 5.5.

A recent paper by Ellis (2006) provides additional information about how to formulate some of the variants to the Araldite, Epon, and Spurr's embedding media.

5.3.2.2 Acrylic Resins

These resins are formed by the polymerization of the monomeric derivatives, generally esters or amides, of acrylic acid. The monomers, some of which are miscible with water, have a very low viscosity, which enables them to quickly infiltrate porous specimens. The resins are transparent colorless polymers (see for example the poly-methylmethacrylate resin Perspex [Lucite]). Unfortunately, acrylic resins have several disadvantages when compared with epoxy resins:

1. The polymerization process of some resins are exothermic and can reach as high as 375 K.
2. The polymerized resins are generally more unstable in the electron beam than the epoxy resins.
3. The stability and mass loss during irradiation in the electron microscope is significantly higher than the losses in epoxy resins.
4. Some of the resin monomers will polymerize in air.
5. Some of the polymerized resins and sections cut from such resins may swell.

In spite of this catalog of perverse problems, some of these resins can be successfully used as embedding materials for scanning electron microscopy. Some of the acrylic polymers are polar and miscible with water and may be used to partially or completely dehydrate wet specimens. Some of these polar resins will even polymerize in the presence of water and produce a hydrophilic polymer. It is appropriate to consider three of these water miscible acrylic resins in order to compare them with the

Table 5.6. Dehydration schedule using glycol methacrylate

Glycol methacrylate dehydration	
Small aldehyde stabilized samples are dehydrated in a graded series of GMA monomer: water mixture at 5, 10, 20, 40, 60, 80% GMA. Time for each change depends on the size and porosity of the specimen.	
Glycol methacrylate (80%)	15 min
Glycol methacrylate (100%)	4 × 15 min changes.
Complete embedding mixture	2 × 1 hour.
Complete embedding for polymerization.	48 hour.

From Hayat, 2000.

three water-immiscible epoxy resins considered earlier. The three different resins to be considered are glycol methacrylates (GMA), which are used at ambient temperatures, and the London resins and Lowicryl resin, which may be used at low temperatures.

GMA is hydroxyethyl methacrylate, a colorless liquid with a viscosity of 51 cPa at 298 K. It is polar and completely miscible in its monomeric form with water, ether, and alcohols. In comparison, in its polymeric form it is practicably insoluble in these solvents. It was first used by Rosenberg et al. (1960) for electron microscopy, and although it later proved less effective for high-resolution studies, it is still effective for embedding soft biological tissues for immunocytochemical studies in the scanning electron microscope. Because of the water miscibility of the monomer, GMA can be used as its own dehydration agent. The embedding schedule is carried out at room temperature and is shown in Table 5.6.

Commercially available GMA may contain varying amounts of hydroquinones that prevent the resin spontaneously polymerizing. This inhibitor combines with oxygen in the air to form a quinone that inhibits the polymerization of the monomer. It is important to remove any hydroquinones from the GMA monomer. Methacrylic acid, ethylene glycol, and ethylene di-methacrylate are other potential impurities that must be removed from the GMA because they all have a deleterious effect on the polymerization of the GMA and its usefulness as an embedding agent. It is important to raise these issues when ordering from suppliers. A suggested embedding mixture and schedule for GMA is given in Tables 5.7 and 5.8.

Gerrits and colleagues have carried out extensive studies on the use of GMA for studies in ambient and low temperature studies on the immunohistochemistry and cytochemistry of biological specimens. The preparation and use of GMA require a lot of attention; the papers by Gerrits et al. (1990, 1991) and Gerrits and Horobin (1996), contain a wealth of practical information on the use of this embedding agent.

London resins, LR White, and LR Gold are acrylic resins with very low viscosity (8 cPa at 298 K). The two London resins differ

Table 5.7. Embedding mixture formulation

Resin Mixture	
Solution A	
GMA containing 200-ppm HME (monomer)	90g
BE (plasticizer)	10g
BP containing 20% water (initiator)	0.5g
Solution B	
PEG 400 (external plasticizer)	15 parts
ND (accelerator)	1 part

GMA = glycol methacrylate, HME = hydroquinone monoethyl ether, BE = 2-butoxyethanol, BP = benzoyl peroxide, PEG = polyethylene glycol, ND = N,N-dimethylamine. The PEG acts as a carrier for the accelerator and softens the final polymerized resin block. From Gerrits et al., 1987.

Table 5.8. Glycol methacrylate embedding schedule

Resin embedding schedule
The infiltration schedule is involved and should be carried out in a fume cupboard and when wearing gloves. The formulation for solutions A and B are given in Table 5.7.
1. Stabilized and dehydrated specimens are infiltrated in a 1:1 mixture of 100% ethanol and Solution A for 2 h and then again in for 18 h.
2. Specimens are transferred either to gelatin capsules or small polyethylene molds fitted with aluminum block holders, which act as heat sinks and filled with 2.5 ml of a 30:1 mixture of Solutions A and B.
3. The filled molds or gelatin capsules are left uncapped for 30 min in an atmosphere of nitrogen at room temperature (293 K) to allow any air bubbles to escape.
4. The embedded specimens are polymerized in 2 h.
5. Polymerization can also take place in UV light (315 nm) for 16–20 h at 277 K.

From Gerrits and van Leeuwen, 1985.

slightly in their formulation. The LR White resin is designed as an effective embedding medium between 258–298 K whereas the LR Gold is effective at temperatures as low as 248 K. Although the resin monomers are insoluble in water, they are miscible with water up to 10–12% by volume. The chemical components of the two embedding media are complicated and not fully known, and it is best to purchase kits of the resins direct from the appropriate suppliers. Like other acrylic resins, the resins may be polymerized by heat, addition of an accelerator, or UV light. The two resins are primarily designed as embedding agents for enzyme histochemistry and immunocytochemistry studies with biological material. The publication by Newman and Hobot (1993) provides additional details of the formulation and use of these two resins. A recent paper by Bowling and Vaughn (2008) has useful information about how to minimize heat damage during thermal polymerization of LR White resin.

LR White's hydrophilic nature allows it to penetrate and fill all the pores and cracks in damp specimens in about 3 h. Polymerization, which is exothermic, is carried out in an anoxic atmosphere at 333 K and the resin has negligible shrinkage and excellent edge retention, and may be polished to a fine surface. A suggested embedding procedure is given in Table 5.9.

LR Gold is similar to LR White but has a different dehydration and embedding procedure, which is shown in Table 5.10.

Lowicryl resins are aliphatic cross-linked acrylate-methylacrylate mixtures, developed by Kellenberger and colleagues in Basel 20 years ago, primarily as embedding agents for low-temperature biological microscopy (Kellenberger et al. 1980; Armbruster et al. 1982; Aceterin et al. 1986; Carlemalm and Villiger 1989).

Table 5.9. LR White embedding procedure

LR White embedding
1. Small stabilized specimens are dehydrated in ethanol at room temperature. Do not use acetone. 70% ethanol　　　　　　　2×15 min 90% ethanol　　　　　　　2×15 min 100% ethanol　　　　　　2×15 min
2. Infiltrate the specimens with the complete LR White mixture in a rotating device in a fume cupboard for 1 hour at 273K.
3. Transfer the samples to small gelatin capsules completely filled with fresh resin and then capped. The loaded capsules are then placed in fitted holes in an aluminium block each of which contain a small amount of ethanol to ensure thermal conduction.
4. Polymerize the resin by placing the loaded capsules in an oven, gently flushed with nitrogen gas, at 333K for a day, or for 1-2 days at 323K.

From Glauert and Lewis, 1998.

Table 5.10. LR Gold embedding procedure

LR Gold embedding
1. Stabilized specimens placed in a 50% MeOH + 20% PVP mixture for 15 min at 277 K.
2. Transfer to 70% MeOH + PVP for 45 min at 253 K.
3. Transfer to 90% MeOH + PVP 2:1 for 45 min at 253 K.
4. Transfer to 90% MeOH + PVP + LR Gold resin mixture at 5:1:5 for 30 min at 253 K.
5. Transfer to 90% MeOH + PVP + LR Gold resin mixture at 3:1: 7 for 60 min at 253 K.
6. LR Gold resin mixture for 60 min at 253 K.
7. LR Gold resin mixture + 5% BME for 60 min at 253 K.
8. LR Gold resin mixture + 5% BME overnight at 253 K.

BME = benzoin methyl ether (inhibitor); MeOH = methyl alcohol; PVP = polyvinylpyrrolidine MW 44 K (provides a colloidal solution and osmotic balance).The embedded tissues are placed in small gelatin capsules, filled to the top with fresh LR Gold resin and tightly sealed to avoid oxygen inhibition of polymerization. The sealed capsules are exposed to UV light at 360 nm for a week at 253 K. From Hayat, 2000.

The books by Griffiths (1993), Chapter 8 of the book by Glauert and Lewis (1998), and Chapter 3 of the book by Hayat (2000), contain additional practical details of how these resins may be formulated and used. They are low-viscosity resins at 298 K but the viscosity increases as the temperature is lowered. In common with most acrylic resins, the Lowicryl resin polymerization is exothermic and susceptible to oxygen and should be polymerized at low temperatures in an anoxic atmosphere.

The resins, which are available from chemical supply companies, are the polar resins K4 M and K11 M and the non-polar resins HM20 and HM23. Depending on the type, the different resins will polymerize between 223 and 192 K. The composition of just one of these resins is described in Table 5.11.

Samples destined for Lowicryl embedding should be prepared by one of the low-temperature stabilization and dehydration procedures discussed in Chapter 7. For example, specimen may be stabilized at room temperature or low temperature and dehydrated at high or low temperatures. Specimens to be embedded by one of the Lowicryl resins may be dehydration by freeze drying, freeze substitution, or the progressive lowering of temperature methods. The starting temperature for embedding is at 273 K or lower. These procedures are best carried out in specialized equipment that allows the processes of dehydration, embedding, and polymerization to be carried out as a continuous process without having to change the container. This type of equipment is discussed in Chapter 7. A dehydration-embedding schedule for one of the Lowicryl resins is shown in Table 5.12.

The low temperature resin embedded samples are polymerized at low temperatures; this procedure is usually carried out in the same specialized equipment used for dehydration and embedding. Glauert and Lewis (1998) give details of different pieces of commercial equipment, and Hunziker and Schenk (1984) describe a simpler device that can be constructed in the laboratory. Table 5.13 provides an example of the polymerization procedure.

Table 5.11. Composition of Lowicryl resin HM20

Chemical	Amount
TEGD (cross-linker)	3.0 g
EM 68.5% + HM 16.6% (monomer)	17.0 g
BM (initiator)	0.1 g

BM = benzoin methacrylate; EM = ethyl methacrylate; NM = n-hexyl methacrylate; TEGD = triethylene glycol dimethacrylate. The embedding mixture should be prepared just before use in a well fitted glass container, mixed using a gentle stream of nitrogen gas and placed in the dark at 253 K. After Glauert and Lewis, 1998.

Table 5.12. Dehydration and embedding procedure for Lowicryl HM20

Dehydration and embedding
1. Pieces of the aqueous stabilized specimens should be placed in small gelatin capsules under liquid and capped. The capsules should fit into small holes in an aluminum block, each of which contains a small amount of ethanol, to ensure thermal conduction.
2. Using a glass Pasteur pipette, remove nearly all the stabilization fluid and immerse the specimen in 30% ethanol at 273 K for 30 min.
3. Remove the 30% ethanol and refill the container with 50% ethanol and leave for 60 min at 253 K.
4. Repeat the process and transfer to 70% ethanol at 238 K for 60 min.
5. Transfer to 95% ethanol at 223 K for 60 min.
6. Transfer to 100% ethanol at 223 K for 60 min.
7. Transfer to 100% ethanol at 223 K for 60 min.
8. Transfer to 100% ethanol:r5esin mixture 1:1 at 223 K for 60 min.
9. Transfer to 100% ethanol:resin mixture 1:2 at 223 K for 60 min.
10. Transfer to pure resin mixture at 223 K for 60 min.
11. Transfer to pure resin mixture at 223 K overnight.

After Villiger 1991.

Table 5.13. A low temperature polymerization procedure for Lowicryl resin

Resin polymerization
1. The polymerization is carried out in a temperature controlled cold box flushed with a gentle flow of dry nitrogen gas to ensure an anoxic anhydrous environment.
2. Depending on the type of equipment, the polymerization may be carried out either in small gelatin capsules or aluminum foil holders.
3. The cold embedded samples are transferred either to OO gelatin capsules or 2-mm deep aluminum foil holders and filled with fresh resin. The gelatin and foil holders should fit into the same aluminum blocks used as heat sinks during embedding.
4. The resin is slowly polymerized over 1–2 days with long wave (360 nm) UV radiation at temperatures between 238–223 K, depending on the type of resin. In some cases the polymerization may be completed at 273 to 288 K.
5. It is important to carefully control the temperature during polymerization and ensure the equipment is dry and free of oxygen until polymerization is complete.

After Glauert and Lewis, 1998.

5.4 Suggested Embedding and Mounting Procedures for the Six Different Types of SEM Specimens

5.4.1 Metals, Alloys, and Metallic Materials

If the samples are only to be imaged and not analyzed and are large enough to be handled manually, they will not need embedding and mounting. If the specimens are to be polished for analysis, they will need embedding and mounting. Care has to be taken with small particles embedded using hot methods, as the high temperatures may cause changes in the microstructure or even oxidation (Goldstein, private communication 2008). It is probably better to use thin layers of epoxy resins.

5.4.2 Hard, Dry, Inorganic Materials

The cold methods can be used with these samples and two examples are given in the following.

5.4.2.1 Rock Samples

1. A diamond saw is used to cut a piece of rock 20 mm long and 10 mm wide and deep.
2. The long face of the cut sample is then mounted on to a standard 25 × 75 mm (ca. 1 × 3 in.) glass slide using a very thin layer of epoxy glue with a refractive index of 1.54.
3. Once the bonding is complete, it may be cut and lapped for microscopy and analysis.

5.4.2.2 Soil Sample

1. The sample is collected in situ by pressing a sufficiently robust open face aluminum box into the soil. The size of the box depends on the size of the soil profile being examined, but typically they should be no bigger than 50 × 30 mm and 100 mm long. This sampling method is used by archaeologists who want to study a profile at a particular location from an excavated site. A metal lip seals the metal box, and care must be taken not to disturb the contents.
2. The sample is dried in an air-assisted oven at 25°C.
3. 1800 ml of the resin and hardener Polylite ABL (www.euroffice.co.uk) is prepared and thoroughly mixed with 200 ml of acetone.
4. The resin mixture is poured over the soil specimen in the aluminum box and kept at a low vacuum of 1.6–3.33 kPa (12–25 mmHg) for 48 h to allow any air bubbles to be removed.
5. The resin is polymerized for 4–6 weeks at room temperature.
6. Finish the polymerization for 24 h at 323 K.

5.4.3 Hard and Firm, Dry Natural Organic Material

This is embedded using cold polymerized epoxy or acrylic resins.

5.4.4 Synthetic Organic Polymer Material

This is embedded using cold polymerized epoxy or acrylic resins.

5.4.5 Biological Organisms and Materials

These specimens make use of the wide range of different epoxy and acrylic resin embedding procedures.

5.4.6 Wet and Liquid Materials

Liquid samples do not need to be embedded, but depending on how much water there is in wet samples they can be embedded in one of the selected hydrophilic acrylic resins.

5.5 The ultimate Destination of Embedded Specimens

As mentioned at the beginning of this chapter, the purpose of embedding is to firmly secure specimens for subsequent microscopy and analysis, but at the expense of completely covering the sample. Depending on the physical nature of the sample and the polymerized embedding material, the circum adjacent specimen may be exposed by grinding, sawing, fracturing, or sectioning. Details of these procedures are discussed in Chapter 6.

6
Sample Exposure

Having selected the sample in terms of its size (Chapter 2), support (Chapter 4), and the ways it may, if necessary, be embedded (Chapter 5), it is now necessary to consider how best to expose the relevant parts of a sample for microscopy and analysis. Scanning electron microscopy is all about examining surfaces. In addition to examining the outside of a sample, it is frequently necessary to obtain additional information about the interior. We may consider surfaces under three categories:

1. The intact natural surface of the specimen
2. The intact natural surface of a discrete internal component of the selected specimen, i.e., a failed engine component, an animal liver, a crystal inside a rock geode, pollen grains inside a flower, ketchup in a hamburger, and the components that make up microelectronics and integrated circuits
3. The general interior of the selected specimen

Intact natural surfaces should need little further exposure, although they require attachment and cleaning. The second category of samples must be opened carefully without damaging the discrete internal surfaces that are to be studied. Most investigations carried out with the SEM are on the internal parts of specimens. One should aim to expose these surfaces using minimal physical force and avoid chemical damage in order to retain the natural features of the sample.

There are many ways to expose the surface of a sample, and it is proposed to first discuss these activities in general terms before considering how best to approach exposing specimens from each of the six different groups of samples. The catalog of methods is considered under four general headings. Methods for use at low temperatures may be found in Chapter 8:

1. The mechanically and sometimes physically violent methods
2. The some what gentler mechanical methods
3. Methods that make use of chemicals
4. Methods that rely on high energy particles

Having successfully exposed the sample, it is necessary to clean the surface. This is discussed in Chapter 10; although specimen exposure and specimen cleaning are conjoined, there is one important difference. It is first necessary to *expose* the specimen before it is *cleaned*. Sample exposure is usually a single process, whereas sample cleaning is continuous and repetitive and carries on up to the point when the sample is placed into the microscope column.

6.1 Rigorous Mechanical and Physical Methods

6.1.1 Exposure by Breaking, Cleaving, Snapping, and Pulling

Breaking involves separating the specimen into parts by the sudden and violent application of force. A hard stone will break into pieces when hit with a hammer; a sharp tool will cleave a plastic dish; a piece of dry wood will snap apart when bent; a metal rod will come apart when pulled with sufficient force. Hard pieces of homogeneous material, such as a rock or a brittle metal, can be ground to smaller particles in an appropriate grinding mill. These violent methods are, literally, hit and miss and because the intact sample is inevitably damaged; one can never be sure that any or all of the pieces of material are representative of the unbroken sample. The only advantage is that it is a very simple way to provide samples of a more manageable size.

6.1.2 Surface Fracturing

This is best described as a more or less random separation of a specimen along a line of least resistance parallel to the applied force. It may be achieved either by applying tensile strength via impact with a blunt instrument, which initiates a single fracture plane, or by using a sharp instrument, which cuts and generates shear forces just ahead of the knife edge to produce a series of conchoidal fractures. Figure 6.1 shows how these two types of fractures may be formed.

Unlike sectioning, the fracturing tool should not make physical contact with the freshly exposed fracture face in order to avoid damaging the undulating and rough surface. This minimal physical contact between the tool and the specimen enables complementary fractures to be made in some types of specimens, as shown in Figure 6.2.

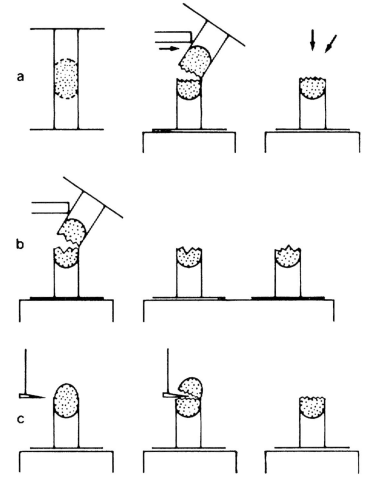

Figure 6.1. Simple and complementary fracturing. A. Impact force to give a simple fracture. B. An impact force to give complementary fractures. C. A knife used to give a simple fracture (Echlin, 1992)

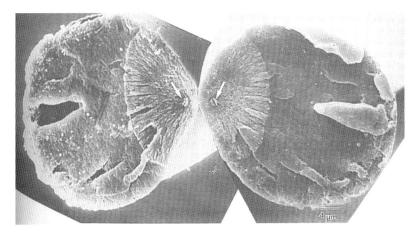

Figure 6.2. A pair of complementary fractures of a polyethylene terephthalate (PET) fiber. The *arrows* show a defect at the locus of the failure (Picture courtesy of Sawyer et al., 2008)

Some metals may be fractured by applying a strong force at opposing points of a sample and literally pulling it apart. An example of this type of fracture is shown in Figure 6.3. Other metals may be impact fractured by first cooling them to very low temperatures, which makes them very brittle. This type of fracture is shown in Figure 6.4. There are fewer surface defects and artifacts on the fracture face, provided the fracturing process is carried out in conditions of low contamination.

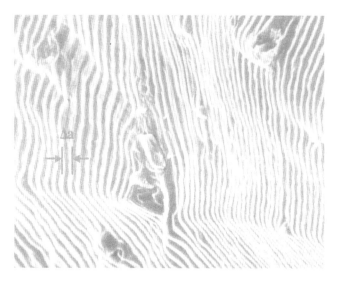

Figure 6.3. Fatigue striation spacing on a face centered cubic (FCC) metal. Each striation represents the incremental advance of the crack associated with one loading cycle (Picture courtesy of Richard Hertzberg, Lehigh University)

Figure 6.4. An ultra-low carbon steel rod cooled to liquid nitrogen temperature (77 K) and impact fractured. The specimen was allowed to return to room temperature before being photographed

The internal structure of flexible organic specimens such as many polymer fibers and films can be exposed using a technique devised by Scott (1959). Scott describes the procedure as an orientation and cleavage plane splitting, which opens the interior of the specimen with minimal disruption. The process is initiated by using a sharp blade to cut halfway into a specimen, such as a polymer fiber, at an oblique angle across the long axis of sample. A second cut is made parallel to the fiber axis. The exposed end of the second cut below the first cut is firmly held with forceps and peeled back by pulling the cut end of the polymer along the fiber axis. Figure 6.5 shows the internal structure of a fiber revealed using this method.

With many chemically complex and soft specimens, the fracture path follows the line of least resistance, but these pathways vary with different materials. It might be along the interface between phases, along regions in which differential contraction has occurred during cooling, or as is now widely accepted, along the hydrophobic inner face of frozen biological membranes that are examined in very high resolution images of freeze-fractured specimen. Freeze fracturing is also a useful procedure for revealing lower resolution images of biological specimen which are to undergo x-ray microanalysis. Low temperature fracturing methods are discussed in more detail in Chapter 8, but Figure 6.6 shows a cross-section of frozen fracture face of a fresh tea leaf.

The process of fracturing is reasonably well understood. Completely brittle multiphase materials, such as many metals and rocks, fracture without deformation, whereas completely ductile materials do not fracture but exhibit plastic deformation.

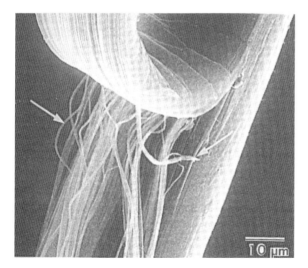

Figure 6.5. An SEM image of the interior of a polyethylene terephthalate fiber prepared by the peel back technique to show the internal fibrillar texture (*arrow*) (Picture courtesy of Linda Sawyer in Goldstein et al., 2004)

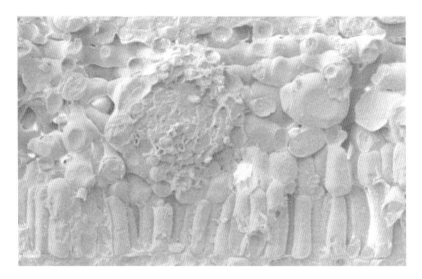

Figure 6.6. Frozen fractured fresh tea leaf at 123 K. Picture width 500 μm (P. Echlin, unpublished image, 1998)

The situation is much less clear in single-phase material, and the fracture pathway may be related to differences in the local orientation of the constituent molecules. Very few organic and hydrated materials are brittle, even at 4 K. Our aim must be to obtain brittle fractures with a minimum of plastic deformation. In an ideal brittle fracture, deformation is limited only to the molecules that are actually pulled apart.

The applied forces of the fracturing tool cause stress to build up in localized regions of the specimen. This stress overcomes the cohesive properties of the specimen and then suddenly spreads rapidly through the sample, causing it to fracture. The way the sample fractures depends on its brittleness, which is influenced not only by the properties of the sample, but also by temperature, rate of stress application, and the presence of discontinuities such as minute cracks in the specimen.

6.1.3 Surface Chipping

Another simple way to expose the surface is to carefully chip away hard material with a sharp tool such as a chisel. The gradual and more controlled chipping process dresses, works, and shapes the surface in the way a sculptor works on a piece of stone. Because only small pieces of material are removed at each percussive hammer blow, the specimen is likely to be less damaged. A bolder approach is to increase the force of the blow and initiate a random fracture through the sample. The one disadvantage of the surface chipping process is that unless you have the skills of Michelangelo, it is not possible to get a very smooth surface.

6.1.4 Sample Cutting

This familiar procedure involves either using a sharp blade or a sharp saw.

6.1.4.1 Blades

These cutting devices can be made from a wide variety of different materials such as diamond, sapphire, ceramics, and glass, and are available in a wide variety of forms. Cutting edges occur in razor blades, kitchen knives, surgical scalpels, and the blades used for cutting thin sections for microscopy and analysis. Knives suitable for preparing samples for SEM are available from most supply houses. Diamond blades may be obtained from www.diatome.ch, sapphire blades from www.wpi-europe.com and ceramic knives from www.agarscientific.com. Blades range in size and shape and need to be held, either in simple holders, surgical holders, or a variety of the different microtomes used to cut thin sections. Figure 6.7 shows a set of disposable scalpel blades available from www.2spi.com. A simple and effective knife can be made by cutting a 2- to 3-mm piece across a thin razor blade using strong scissors and placing it in a suitable holder.

Cutting Thin Sections at Ambient Temperatures
Any cutting action produces two faces. If the cutting action is designed to produce a thin sliver of the sample, we refer to this as a *section*; many pieces of equipment are designed to produce thin sections. Thin sections of material are a good way to expose the interior of a sample. The material from which the section was cut is referred to as the *block face*. It is somewhat more usual to examine the block face of the specimen rather than the section cut from this face, and many pieces of equipment are used to *plane* the surface of the specimen and discard the cut sections.

During specimen cutting the separation of the sample occurs at the knife edge itself, and in this respect the process of sample separation involves localized compression, which may damage the specimen. This damage can be minimized by using very sharp blades. The actual molecular and physical processes of cutting are not fully understood. It is not clear whether the sharp

Figure 6.7. Disposable scalpel blades which may be used for cutting small soft specimens (Picture courtesy of SPI Supplies USA)

knife makes continuous physical contact with the specimen or is the result of continuous fracturing.

The extensive literature on metal machining suggests that soft metals are cut as a continuous chip, whereas hard metals are cut as a discontinuous chip. Kellenberger et al. (1985) and Aceterin et al. (1987) on the basis of extensive work with soft resin-embedded biological materials, consider that both fracturing and cutting at ambient temperatures are cleavage phenomena in which material is pulled apart. As material is cleaved, it first passes through a plastic phase in which the material elongates in a reversible manner before undergoing irreversible plastic flow. The plastic flow continues until the material ruptures, after which the elastic elongation may be reversed, but the plastic deformation remains unchanged. The relative amounts of elastic elongation and plastic flow vary according to temperature, the physical and chemical composition of the material, and the strength and toughness of the cutting tool. Thus a fine steel, ceramic, or glass knife cuts most soft biological materials and many plastics. A diamond or a sapphire knife, in addition to cutting soft materials, also cuts some soft metals. There are no knives that cut hard inorganic and organic material.

Cutting can be achieved either by moving the blade over a stationary specimen or moving the specimen over a stationary blade. The rotating blade in a meat slicer, the fine stainless steel knife cutting the Christmas turkey, and a surgical scalpel are best used by slowly drawing the blade down and toward across the sample. This type of cutting process causes the least compression damage to the sample.

Planning Block Surfaces
The alternative cutting process of moving the specimen across a stationary blade also can be used in microtomes and ultramicrotomes but discards the thin sections and uses the smooth specimen block that remains behind. This process is referred to as planing. The remaining freshly cut surface can be examined in the SEM. An alternative to mechanical planing is to use high energy particles to selectively erode away parts or all of the specimen's surface. These procedures are discussed in section 6.4.

The advantages and disadvantages of examining the block face rather than the thin section produced by cutting are discussed by Walter (2003) and Richter et al. (2007) on preparing material for low temperature microscopy. They found that high pressure frozen specimen blocks, cryo-planed at 133 K with a diamond knife, did not exhibit any of the usual artifacts normally associated with cryo-sections.

In all cutting processes, the least damage occurs when very sharp thin knives are used to cut soft material and the potential for damage increases when dull knives are used on increasingly hard samples. The end product of a sectioned or planned surface is generally much smoother than a surface which has been revealed by sawing.

6.1.4.2 Saws

In the context of this book, saws are used to expose hard materials such as metals, rocks, and some hard organic materials. Saws range in size from small jewelers' saw to saws in timber mills. The saw blades have a serrated edge, which either may be pushed backward and forward across the stationary specimen or moved continually past the stationary sample using a powered blade. Unlike a blade, the cutting action of a saw completely removes part of the specimen and damages cut surfaces. For this reason, saws should only be used to cut a large specimen down to size and not to prepare a surface for subsequent imaging and analysis.

The blade of a saw used for cutting wood is made of steel. Diamond saws are used to cut harder material such as metals and rocks. Diamond saws are made by combining powdered metal with diamond crystals, which are then heated under pressure in a mold to impregnate the diamond segments into the metal. Diamond saws are manufactured as continuous rim discs or as wires that are rotated past the sample in a controlled manner. The most familiar form of diamond saws are those used by road engineers to cut and shape paving stones. The saws used in laboratories are a little more sophisticated and are available from a number of different companies. For example, the company Logitech in Scotland (www.logitech.uk.com) makes the thin section cutoff and trim saw, shown in Figure 6.8, which is suitable for cutting geological specimens prior to the lapping procedure discussed in the following.

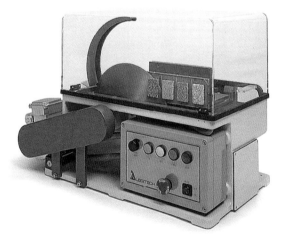

Figure 6.8. A thin section cut-off and trim rotary saw for cutting specimen from geological samples (Picture courtesy of Logitech Ltd., Glasgow, Scotland, UK)

6.1.5 Lapping

This is a mechanical operation in which two surfaces, separated by an abrasive material, are rubbed against each other. The rubbing motion either can be by hand or by using a machine. Lapping is an important preparative procedure for the examination of metals and hard inorganic materials, and a large number of companies make the appropriate equipment. Lapping is composed of two sequential processes; *grinding* and *polishing*. Figure 6.9 (courtesy of Joe Goldstein) is part of a chondrite (a stony meteorite) that has been cut to shape, mounted in plastic, and then lapped prior to examination and analysis.

6.1.5.1 Grinding

Mechanical Grinding

This is a very popular process and involves rubbing brittle and/or hard samples against a surface such as iron or glass, referred to as the lap or grinding tool. Abrasive material is placed between the two surfaces. This arrangement produces microscopic conchoidal fractures as the abrasive material moves between the two surfaces. Grinding is carried out using a mechanically rotating flat sheet of abrasive material against which the sample is either held by hand or moved by machine. The grinding motion progressively removes both the sample surface and the abrasive lap.

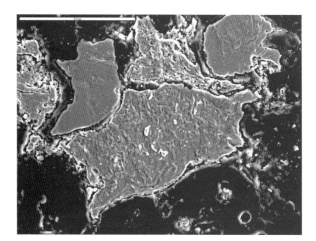

Figure 6.9. Microstructure of unzoned metal particle in the Northwest Africa (NWA) 739 CH chondrite. SE photomicrograph of two unzoned particles that are in intimate contact. Several particles are plastically deformed as shown by the twisted grains in the microstructure. The sample is mounted in plastic and coated with carbon. Marker = 50 µm (Picture courtesy of Joseph Goldstein. For details see Goldstein et al., 2007)

Electrolytic Grinding
A metal bonded diamond impregnated grinding wheel is used to remove material from a metal specimen. In addition to mechanical abrasion, there is an insulator between the grinding wheel and the specimen that acts as an anode and allows electrochemical machining to occur. The Electromet 4, made by Buehler Ltd. (www.buehler.com), is suitable for preparing metal specimens.

Mechanical Polishing
This process involves using a soft material containing abrasives as the lap against which the harder sample is rubbed. The abrasive material embedded in the soft lap scores across and cuts the harder material. By using progressively smaller and smaller abrasive particles, the sample becomes increasingly more highly polished. The sample can remain stationary and the soft material containing the fine abrasive moves against it in the way we polish our cars. Alternatively, the soft abrasive material may be firmly held and rotated and the specimen held against it manually or mechanically. A large number of different machines are available for both grinding and polishing samples for microscopy and analysis. For example: Buehler (www.buehler.com), Electron Microscopy Supplies (www.emsdiasum.com), Engis (www.engis.com), and Kemet (www.kemet.co.uk). Make the appropriate equipment. Figure 6.10 is the bench-top Engis AM-15 Lapping and polishing machine that is suitable for preparing metals, geological specimens, semiconductors, ceramics, and electron-optical specimens.

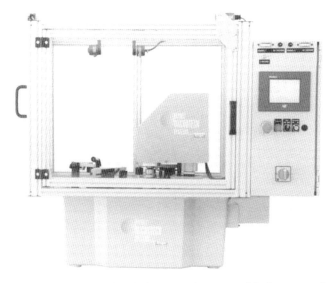

Figure 6.10. A lapping and polishing machine suitable for preparing geological samples, ceramics, semiconductors, and electro-optical specimens for microscopy and analysis (Picture courtesy of Engis Co., USA)

Grinding and polishing are very effective ways of exposing surfaces, and their only disadvantage is that the processes can cause both mechanical and thermal damage to the sample.

Both grinding and polishing depend on the use of either natural or artificial abrasive materials. Natural abrasives include corundum, a very hard mineral oxide of aluminum (Al_2O_3), emery, a fine-grained impure variety of corundum, crushed garnet, a group of silicate minerals containing varying amounts of Mg, Al, Ca, and Fe, quartz (SiO_2), diamond, and pumice, a complex silicate containing Al, K, and Na. These and other minerals have a range of hardness, with diamond the hardest material (10 on the Mohs scale), corundum 9, garnet 8, quartz 7, felspar 6, apatite 5, fluorite 4, calcite 3, gypsum 2, and talc 1. Artificial abrasives, which have hardness between 9 and 10, include alumina, boron carbide, boron nitride, silicon carbide, and synthetic diamond.

The abrasive materials can be supplied as impregnated papers and cloths, as powders, or as fine aqueous suspensions. The abrasives used for preparing specimens for both SEM and microanalysis come in a number of standard grain sizes, as shown in Table 6.1. The website www.cs.rochester.edu gives more details.

Diamond powders and slurries range in size from 25 nm to 80 µm, but usually only five grades (9, 6, 3, 1, and 0.25 µm) are used for polishing samples for microscopy and analysis. The Advanced Abrasives Corp. (www.advancedabrasives.com) has information about these materials.

Although nearly all abrasives contain some impurities, manufactured materials are preferred because the particle size and chemical composition are more uniform and available in different grades. The choice of abrasive is governed by the hardness of material and hardness of the abrasive. The abrasives may be used as granules or powders bonded into metal wheels or bonded into papers and cloth. For example, emery paper or cloth has a surface

Table 6.1. United States and European equivalent grit guide and their actual grain size

U.S.A.	Europe	Grain size (µm)
	P4000	5
1200	P2500	10
600	P1200	15
500	P1000	
400	P800	
	P600	25
320	P400	35
	P320	45
220	P240	60
180	P180	75
120	P120	120
80	P80	200
60	P60	270

of a mixture of natural crystalline corundum with iron oxides with a hardness of between 7 and 9 and is available in a range of compositions and hardness. Although the different abrasives are very effective in removing surface material from hard specimens, they also leave remnants of material behind that contaminate the sample. It is important to constantly clean the sample surface during polishing and after the process has been complete.

The following sequence of events shows how a typical grinding and polishing schedule may be carried out on a geological sample cut from a piece of rock:

1. A diamond saw is used to cut a piece of rock 20 mm, 10 mm wide, and 10 mm deep.
2. The long face of the cut sample is mounted on a standard 75 × 25 mm (ca. 1 × 3 in.) glass slide using a very thin layer of epoxy glue with a refractive index of 1.54.
3. The firmly attached piece of rock is then made thinner by an additional cut with the diamond saw and then lapped to a thickness of 100–30 µm, washed in an ultrasonic water bath, and dried with a jet of clean air.

Step 1: Silicon Carbide Paper. The sample is held by hand and moved in a circular motion on the following sequence of flat graded papers: P240 (60 µm), P400 (35 µm), P600 (25 µm), P800 (22 µm), and finally P1200 (15 µm). The surface of the sample should be cleaned with an air brush between each grade of carbide paper.

Step 2: Silicon Carbide Powder. A small amount of an aqueous suspension of P600 (25-µm) silicon carbide powder is spread on the surface of a firm flat glass plate. The specimen is held firmly and moved randomly in all directions over the whole wet surface of the glass plate until a sheen appears on the specimen when held up to the light after washing and drying. The sample is rinsed in water and then washed in an ultrasonic bath two or three times and then dried with an air brush. The water in the ultrasonic bath must be changed with every stage of the process. The surface is examined in a reflected light microscope and if there is grit in any holes in the sample, then rinse and re-wash in the ultrasonic bath. Clean off the grit suspension from the glass plate.

Repeat the process using a P800 (22-µm) and then a P1000 (18-µm) aqueous suspension of silicon carbide powder. The sample should be washed, ultrasonicated, and cleaned each time. Clean off the grit suspension from the glass plate. The process is finished when the surface, examined through a light-reflecting microscope, has light speckles surrounded by dark patches of vesicles.

Step 3: Polishing Machine. The flat surface of the polishing machine should be covered with a hard, non-woven fiber resin lapping fabric. A small amount of a 9-µm diamond suspension and distilled water is sprinkled on the surface of the lapping fabric and the specimen is held firmly and regularly twisted around in the fingers and moved randomly over the whole moving lapping surface for about 5 min. The sample is then rinsed and washed in an ultrasonic bath two or three times and dried with an air brush. The dry surface is then examined in the light reflecting microscope to see if there are light/dark patches in the image.

Repeat the process with 6-, 3-, and 1-µm and if necessary, 0.5-µm diamond suspensions until the sample surface appears in the reflecting light microscope to be predominantly light with just a few dark-appearing vesicles.

This polishing procedure (provided courtesy of the Department of Earth Sciences, University of Cambridge), is designed to prepare sample for microscopy and x-ray microanalysis. The protocol described above can also be varied for metals, ceramics, and other hard dry inorganic and biological materials such as bones and teeth.

Electro Polishing

This is the electrolytic removal of metal in a highly ionic solution by means of an electrical potential and current. It removes a very thin layer of material on the surface of the metal. The amount of charge applied to the metal is highly dependent on the metal sample and under certain conditions there is a preferential solution so any microscopic irregularities disappear leaving a very smooth surface. There are a large number of manufacturers of suitable equipment including Electro Shine (www.electroshine.com.au) and Electro Polish Systems Inc. (www.ep-systems.com).

The four rigorous mechanical and physical methods described above are the most commonly used procedures to expose and polish most types of sample for the SEM. The following outlines a number of specialized procedures that may be used to provide both chemical and structural information, particularly for metallic and hard inorganic specimens.

6.2 Gentler Mechanical, Physical, and Chemical Methods

6.2.1 Gas-Borne Particle Abrasion

Silica, silicon carbide, or alumina particles are carried in a stream of compresses air at 10.7 kPa (50 psi) or gas from a bottle of argon or nitrogen. A compromise has to be struck between using small (0.5-µm) particles that have a slow erosion rate but give a smooth surface and using larger (30-µm) particles that have a faster

Figure 6.11. The fine sand feed on a gas jet erosion system. The high pressure gas jet is to the *left* of the picture (Picture courtesy of Phoenix Tribology Ltd., USA)

erosion rate but leave a much rougher surface. The abrasion rate is faster and less controlled with softer specimens, and there is an additional disadvantage that the abrading particles can become embedded in the specimen. The surface finish can be as good as 200–300 nm, but this is dependent on the size and composition of the abrading particles. There are a number of different suppliers of equipment; Figure 6.11 (Phoenix Tribology Ltd., www.phoenix-tribology.com) provides a picture of the type of equipment required for the erosion process.

6.2.2 Mechanical Thinning

Goodhew (1973) showed that for most metals and hard dry inorganic specimens, purely mechanical techniques such as rolling, cutting, and grinding do not produce a sample thin enough for imaging with transmitted electrons. Mechanical thinning is used to reduce the depth and thickness of the specimen by a process usually referred to as dimpling. This uses fine wet grit abrasives and involves a vertical rotating grinding wheel that oscillates horizontally to give a central thin layer on the sample. The specialized equipment needed for this process is available from a number of different companies. Figure 6.12 shows the Model 200 Dimpling Grinder available from Fischione Instruments (Export, PA, www.fischione.com).

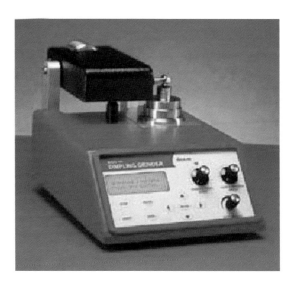

Figure 6.12. A dimpling grinder for thinning metallic specimens to a few micrometers thick prior to ion beam milling (Picture courtesy of Fischione Instruments, Pittsburgh, PA)

6.3 Methods That Make Use of Chemicals

It may be necessary to follow the two mechanical procedures discussed in the preceding by chemical methods and techniques that make use of high energy beams. These procedures are discussed in the following.

6.3.1 Chemical Thinning

Although mechanical grinding and polishing techniques are very effective in exposing surfaces, they may cause changes in the surface's ultrastructure. This is a particular problem in materials composed of many different hard and soft phases or in which there are distinct boundaries between areas of different composition. Chemical thinning can involve no more than simply dipping the specimen into a static or gently agitated solution, spraying the specimen with a gentle stream from a rubber pipette or with a more powerful jet stream.

Difficulties begin to arise with metallic specimens that have surface layers of oxide and corrosion. The etching and chemical treatment can alter the chemical composition of the metal surface and create height differences between different phases of the metal. Care has to be taken with passivation techniques that use chemical agents to make a polished metal surface inactive or less active to corrosion or use de-passivation techniques that remove existing oxide layers and allow new oxide layers to form. Unlike polishing and ion beam milling methods (see later), the chemical etching processes are designed to reveal particular topographic substructures such as rystal structures, grain boundaries, and changes in composition. Chemical etching agents are developed by trial and error and the book by Van der Voort (1984) and the various handbooks by the American Society of Metals contain extensive recipes.

6.3.2 Chemical Polishing

A wide range of chemical recipes have been developed for metals and more difficult samples such as semiconductors, glasses, some metamorphic rocks, and gemstones such as garnet. The book by Goodhew (1973) contains 15 pages of chemical recipes that, although developed for transmission electron microscopy, are applicable for use with the SEM. The simplest technique is to hold the sample with a pair of tweezers and immerse it half way into a bath of stationary polishing solution for increasing periods of time. This allows one to follow the cleaning process. More recalcitrant specimens may need to be exposed to moving solutions or jets of liquid. The Goodhew reference (1973) is full of practical suggestions. Once the polishing process is finished it is most important to thoroughly clean the specimen two or three times with a clean solvent followed with two rinses in either methanol or ethanol.

6.3.3 Electrochemical Polishing

The mechanism is very similar to chemical polishing described in the preceding except that the addition of an electric potential makes the procedure more controllable. The book by Goodhew (1985) contains much useful and practical information about this method, including an extensive 12-page list of different recipes for chemical and electrochemical polishing solutions for a wide range of metals, alloys, and ceramics.

Electropolishing is electroplating in reverse. The specimen is placed into a polishing bath containing an electrolyte, a source of applied potential, a meter to measure the current, a cathode, and a stirrer. Goodhew (1985) explains that during electropolishing at least two layers should be encouraged to form on the specimen's surface. A relatively thick viscous layer that controls smoothing any macromolecular irregularities and a thinner layer below promotes a highly polished surface. The stability of these two layers is controlled by the chemical composition, temperature, and movement of the electrolyte, the applied potential, and even the orientation of specimen in relation to the cathode and ground. Once the polishing process is finished it is most important to thoroughly clean the specimen two or three times with a clean solvent followed by two rinses in either methanol or ethanol.

The chemical and electrolytic etching/thinning methods have two effects. The reactive agents may be chosen either to remove specific components from the specimen surface or smooth surfaces that show surface irregularities after polishing. The sample surface should be split into two closely identical parts. Powerful oxidizing agents are fused to oxidize the surface of one part of the specimen's surface. The oxidized surface film is then removed using another agent to allow the first agent to continue the oxidation process. The second agent is frequently referred to as a depassivator, which has the effect of preventing the normal electropotential of the sample surface from being masked. Finally, a viscous component is often included to produce a thick surface layer for macroscopic polishing. The ultimate effect is to produce a chemically etched specimen with a relatively thick layer that controls the smoothing of irregularities. The treated and untreated surfaces may be polished to a fine degree. Care must be taken to remove both the etching and polishing materials, particularly if the sample is to be analyzed by x-ray microanalysis. Full details of these procedures are to be found in the excellent book by Goodhew (1973).

6.3.4 Surface Replicas and Corrosion Casts

These are specimen exposure mechanisms that may be used to expose very long tortuous holes or very long naturally exposed surfaces. Casts of fracture faces or exposed surfaces can be made using dental wax, silicone rubber, and various plastic films, which are then given a conductive coat and examined in the microscope.

Corrosion casts work on the principle of making a cast of a space, for example a blood vessel, using latex, methacrylate, or polyester resins, and discarding the surrounding solid material. The same techniques can be used for other biological spaces such the digestive tract, excretion pathways, and internal airways. In the case of biological material, the space is first rinsed clean by diffusion with a suitable solution, or in the case of biological material, a suitable buffer and the space is filled with a liquid resin such as Mercox and left in situ to polymerize. Careful mechanical dissection is followed by appropriate chemical stabilization (see Chapter 8) after which the remaining tissue that surrounded the space is removed by maceration in 25% NaOH.

Corrosion casts are not readily applicable to the other five general types of specimen, for although the resins can be infiltrated into internal spaces it is difficult to expose the polymer casts by corrosion methods that do not damage the cast itself. For example, it would be difficult to remove a polymer cast from inside of a convoluted metal pipe.

The papers by Hirschberg et al. (1999), Sims and Albrecht (1999), and more recently Stöttinger et al. (2006) review these procedures for microvascular corrosion casting. The chemicals, equipment, and methods for corrosion casting can be found at www.2spi.com and www.tedpella.com.

6.4 Methods That Rely on High Energy Particles

There are several different ways high energy beams may be used to expose and clean specimens either by using low energy gas plasma or a low energy beam of ions. The different methods can influence the sample surface in three ways:

a. *All* of the sample surface
b. Selected *parts* of the sample surface
c. *Clean* contamination from the sample surface

There are two relatively new techniques have given rise to very effective instrumentation that is of particular use in the semiconductor industry. Many of these systems combine both plasma etching and ion beam etching in the same instrument.

6.4.1 Plasma Etching

This method inductively couples a radio frequency (RF) generator with a low pressure mixture of various gases such as oxygen and argon. The activated low energy gas plasma induces chemical reactions with various components of the sample, particularly hydrocarbons. For example, disassociated oxygen atoms created by the plasma, chemically react with hydrocarbon materials and convert them to CO, CO_2, and H_2O that are removed by the vacuum system. As shown in a later section, oxygen plasma etching is a very effective way of removing hydrocarbon contamination from hard materials.

Other gases may be used to remove other materials or selectively etch microstructural constituents (Isabell et al., 1999). The time needed for etching depends on the amount of contaminating hydrocarbons and the relative reactivity between the plasma gas and the sample. Plasma etching devices are available from many different companies, including Emitech (www.emitech.co.uk), Quorum Technologies (www.quorumtech.com), Diener Electronic GmbH (www.plasma.de), and Fischione Instruments (www.fischione.com). Figure 6.13 shows the K-1050X Plasma etcher/plasma asher from Quorum Technologies (www.quorumtech.com) that is the type of equipment that may be used to expose and clean specimen prior to SEM and x-ray microanalysis.

Plasma etching is primarily a cleaning process that effectively removes hydrocarbons from the surface of metals and hard, dry inorganic specimens. The technique is not useful for exposing the unsullied surface of organic specimens, as one can never be sure when the surface *cleaning* ends and the natural surface begins to *erode*.

6.4.2 Ion Beam Etching

There are two general approaches to ion beam etching. The etching can be carried out using a small dedicated device, a so-called ion beam gun, which uses a poorly focused beam of positive argon

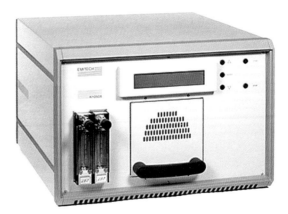

Figure 6.13. Plasma etcher used for cleaning and removing organic material from sample to be examined in the SEM (Picture courtesy of Quorum Technologies, Ringmer, UK)

ions. When needed, the ion beam gun can be fitted to the SEM via one of specimen chamber ports. The more modern approach is to use a dedicated focused ion beam (FIB) microscope, which has a built-in device that rasters a fine beam of positively charged ions over the sample surface.

Before discussing these two different approaches it might be helpful to first consider the advantages and disadvantages of this exposure mechanism. Unlike fracturing, sawing, and cutting, the focused ion beam processes have an increased precision regarding where any cuts—or milling processes as it is referred to—are to be made on the sample. This precision depends on the degree to which the ion beam can be focused. Any material can be milled regardless of its hardness and chemical composition, and the general rule is that the harder the material the longer it takes to mill. Advantage of ion beam etching/milling on hard non-organic samples is that it can remove the damaged layer produced by prior mechanical grinding and give a highly polished surface.

As demonstrated later, ion beam milling plays a pivotal role in the study of metallic samples, and the inspection, imaging, and sometimes the repair of microelectronic devices and semiconductor circuits. Interpretation of the changes in the images of milled samples, although not readily appreciated, can be understood once the precise physiochemistry of the particular ion beam parameters are appreciated. We will return to this later.

The technique is much less useful for soft materials such as polymers, dry organic materials, and biological samples, because the milling process develops heat, removes organic material much faster than inorganic material, and in some cases may contaminate the specimen. A consequence of these adverse effects makes image interpretation rather difficult.

6.4.2.1 Ion Beam Guns

The earliest attempts at investigating the topography of ion beam bombarded surfaces was made in 1962 by Gary Stewart at the Engineering Department, University of Cambridge using a 150-µm radius argon ion beam at 5 kV and total ion current between 10–100 µA. Details of these experiments together with a large number of papers covering the development of the scanning electron microscope in Cambridge, are to be found in the recent book edited by Breton et al. (2004). Alan Boyde together with Gary Stewart published a number of papers in the early 1960s that used an early SEM fitted with an ion gun to study a kangaroo tooth, which, to use Alan Boyd's phrase, "was hammered with an argon beam." This was a prophetic turn of phrase!

Figure 6.14 shows an attempt by the author together with the Cambridge Instrument Company about 35 years ago, to investigate

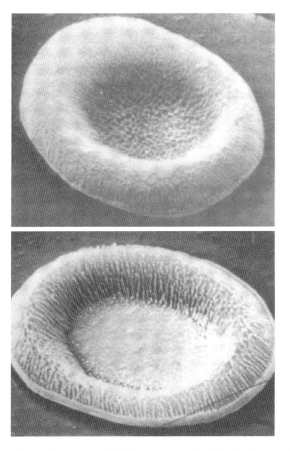

Figure 6.14. *Upper image.* An SEM image at 20 kV of a freeze-dried red blood cell that had been ion beam etched for 30 s. *Lower image.* The same red blood cell that had been ion beam etched for an additional 2 min. Images sputter coated with 15 nm of gold Picture width 6 µm (P. Echlin, 1998)

the usefulness of ion beam etching as a tool to be used to study soft biological tissue in the SEM. As the images show there is no doubt that the ion beam removed material from soft specimens, but the interpretation of the results remains problematic.

Detachable ion beam guns have evolved into three different types of ion beam milling devices. There are the dedicated devices that have an ion source with or without an external stereo light microscope, focused ion beam microscopes, and dual beam devices that have a separate ion source and electron beam source on the same microscope column.

Dedicated Ion Milling Instruments
The most usual way now is to use a narrowly focused beam of positive argon ions to erode away the surface of the specimen. The rate of sample removal is much slower than that achieved with the high energy of the focused ion beam instruments discussed in the following. The etching profile varies according to the tilt angle between the sample and incident ion beam, the beam energy, and the chemical composition of the specimen. The tilted sample is rotated in respect to the incident ion beam, and surface material is removed by a concentration of momentum transfer and/or chemical reactivity. At low angle (15°) incidence to the surface, ions tend to level the surface rather than etch; above 15° the incident ions tend to raise the topography or decorate the microstructural features. Ion beam milling is an important specimen preparation procedure for hard materials such as metals and dry inorganic specimens because it removes the damaged layer produced by mechanical grinding and provides a highly polished surface.

An alternative to ion beam etching is to use reactive ions of different materials to selectively remove different chemical components of the specimen. For example: reactive ion beam etching (RIBE) uses gases such as Cl_2, N_2 BCl_3, and CCl_3F_3 to selectively remove different chemical components of the sample. Reactive ion etching (RIE) uses SF_6 or CF_4 gas in which plasma is formed by an RF source to break the gas molecules into reactive ions. These and other related procedures are much more selective in the way they erode different components of the sample.

Reactive ion etching procedures are used in material science and in the study and manufacture of semiconductors and integrated circuits. A detailed discussion of these procedures is to be found in the recent book by Lieberman and Lichlerberg (2005). A number of different companies manufacture stand-alone ion beam thinning and etching devices, for example, JEOL Ltd. (www.jeol.com) and Bal-Tech RMC (www.bal-tech.com).

The ion beam milling processes of soft organic samples are not fully understood. The general view is succinctly summarized in

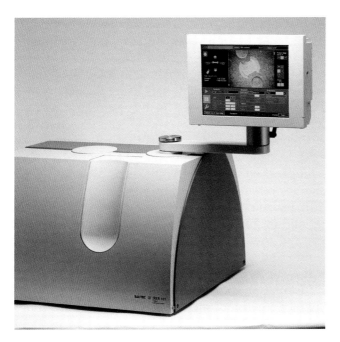

Figure 6.15. The Bal-Tec RES 101 bench top ion beam device for milling and cleaning specimens for SEM, TEM, and light microscopy (Picture courtesy of Bal-Tec AG, Balzers, Lichtenstein)

a recent paper by Boyde (2004), who pointed out that the surface topography of an ion beam milled sample is determined by what has been *removed* from the sample, *not* by what remains.

A number of different companies make ion beam milling instruments; Figure 6.15 shows the RES-101 ion beam milling machine made by Bal-Tech AB (www.bal-tec.com). It uses an argon ion source between 1–10 kV, 200 mA current; and the FWHM of the beam is between 0.8–2.5 mm. This machine is used to mill, enhance contrast, and clean hard samples such as metals, ceramics, glasses, multi-layers and semiconductors, minerals, and bones prior to examination in SEM, TEM, and light microscope investigations. A version of this milling machine can be fitted with a high resolution stereolight microscope in order to study the milling and cleaning processes.

Focused Ion Beam Instrument (FIB)
This type of instrumentation has extended the effectiveness of ion beam milling. A typical focused ion beam instrument uses a gallium liquid-metal ion source operating at 2–30 kV, 1 pA–20 nA beam current, a minimal spot size of 5 nm, and a spatial resolution of 5–7 nm. The very narrow ion beam is rastered across the sample surface and gives rise to secondary and backscattered electrons, ions, photons, and neutrons that can be used to micro-machine, image, and analyze the sample.

FIB instruments are usually equipped with devices to inject different gases at the ion beam impact point, making it possible to selectively etch a fine section across a sample or over the surface. In addition, it is also possible to selectively deposit both conductive and insulating materials using the reactive ion beam etching mechanisms discussed in the preceding. For example, mixed gases of CF_4, CHF_3, Cl_2, and O_2 can give rise to reactive fluorine, chlorine, and oxygen ions that influence Si, SiO_2, and Al. The FIB instruments are very useful for both microscopy and sample preparation of virtually any type of sample from some soft materials to hard rocks and metals.

The FIB instruments play a pivotal role in the production, cleaning, and examination of microelectronics, semiconductors, ceramics, and metals. Recent studies show how a FIB can be used to mill and sharpen, from site-specific regions, extremely fine needle-shaped atom probes.

More information about the instrumentation and wide range of applications of field ion microscopes can be found from the FEI Co. (www.fei.com) Tescan s.r.o (www.tescan.com) and Carl Zeiss NTS GmbH (www.smt.zeiss.com). Figure 6.16 is of a focused ion beam instrument made by JEOL Ltd. (www.jeol.com).

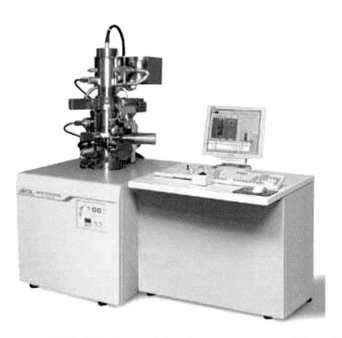

Figure 6.16. JEOL-9320 Focused ion beam system used for milling, polishing, enhanced etching, and deposition for structural and material analysis of semiconductors and hard materials (Picture courtesy of JEOL Ltd., Tokyo, Japan)

Dual Secondary Electron and Ion Beam Instruments

These newer instruments are an obvious extension of the earlier FIB instruments. The dual beam microscope combines the separate columns of a high-resolution (1–3 nm) scanning electron microscope and a high-resolution (3–5 nm) focused ion beam microscope into one instrument. The dual beam instrument permits sample preparation, imaging, and analysis with the FIB and high resolution imaging and analysis with the SEM. The SEM columns operate at between 0.2–30 kV, 1 pA–65 nA and a spatial resolution of 1–3 nm. The FIB columns operate at between 2–30 kV, 2 pA–20 nA, and a spatial resolution of 3–5 nm.

Because it is also possible to inject different gases at the ion beam impact point, it allows it to selectively etch a cross section or the sample surface of a specimen. In addition, it is also possible to selectively deposit both conductive and insulating materials. For example, mixed gases of CF_4, CHF_3, Cl_2, and O_2 can give rise to reactive fluorine, chlorine, and oxygen ions that influence Si, SiO_2, and Al. The dual secondary electron and ion beam microscope plays a pivotal role in the production, cleaning, and examination of microelectronic, semiconductors, ceramics, and metals. More information about these instruments and the wide range of applications can be found from the FEI Co. (www.fei.com) and Carl Zeiss NTS GmbH (www.smt.zeiss.com). Figure 6.17 is of the Quanta 3D FEG made by FEI that has been designed to examine

Figure 6.17. An example of a dual beam microscope—the Quanta 3D FEG, made by the FEI Co.—is a high resolution secondary electron and ion beam instrument that can operate at high and low voltages and low pressure in the SEM mode (Image courtesy of the FEI Company, Hillsborough, OR)

and analyze a wide range of different specimens, including biological and wet specimens.

6.4.2.2 Recent Advances in the Use of Dual Beam Microscopy

One of the most encouraging advances in dual beam microscopy is the recent experiments using these instruments to study soft samples, biological material, and normally wet specimens. Marko et al. (2006) investigated the feasibility of using a focused ion beam to thin vitreously frozen biological specimens. Concern had long been expressed that heat transfer below the direct ion beam interaction layer might damage delicate samples and, in this particular case, devitrify the ice. Marko and colleagues found no evidence of heat-induced devitrification after focused ion beam milling through a depth of 1.0 mm into the specimen.

A more recent study by McGeoch (2007) used a cold stage dual beam instrument (SEM + FIB) to study the topology of pieces of mammalian material quench frozen in liquid nitrogen. The samples were maintained at 123 K and a system of top-down focused ion beam milling was used as a "gentle chisel" at between 10–50 pA and 30 kV to efficiently etch samples. Although this is a promising start, the results are difficult to interpret because it has been known for some time that top-down focused ion beam milling creates many artifacts because of preferential milling. Another problem in interpreting the results is that the author assumed the samples were vitrified. This is most unlikely in specimens that were quench cooled in liquid nitrogen (see later in Chapter 7).

Hayles et al. (2007) extended these pioneering studies, and in addition to confirming the earlier work also devised a method to deposit a protective flat surface metal layer onto a frozen sample, enabling high quality cross-sectioning using the focused ion beam.

Small pieces of a tobacco flower petal were quench frozen in melting nitrogen at 63 K and immediately transferred, via an air lock, to a low temperature specimen preparation device attached to the side of the microscope and kept at 148 K. The frozen specimen surface was sublimed at 178 K for 2 min to remove any surface ice that occurred naturally on the flower petal. The samples were cooled to 148 K and sputter coated with 15 nm of platinum and then transferred, via a second air lock, to the cold stage of the dual beam microscope.

The focused ion beam was used with a gas injection system to deposit 1–2 μm of additional platinum on to the cold specimen for protection from the beam during milling by means of gas vapor deposition. The surface of the now doubled platinum coated cold specimen was milled and polished with the focused ion beam at 30 kV and 20 nA. The resulting section milled into the sample, was sublimed at 178 K while inside the microscope chamber to enable the slow sublimation process to be monitored visually. The specimen was then withdrawn back into the adjacent low temperature specimen preparation unit and sputter coated with an additional 2 nm of platinum.

A summary of the sequence of this procedure is as follows:

a. The clean frozen fractured sample surface was first sputter coated with a thin layer of metal to make it *conductive*.
b. The conductive specimen was then *protected* with a relatively thick flat metal surface by chemical vapor deposition using the focused ion beam of the microscope.
c. The focused ion beam was then used to *mill and polish* selected regions of the frozen specimen.
d. The resulting exposed region of the specimen was sublimed at 178 K.
e. The newly exposed regions of the sample were given a second very thin layer of *conductive* metal.
f. The frozen specimen was *imaged* using the secondary electron beam of the microscope.

Figure 6.18 is from the work of Hayles and colleagues and shows part of the surface of a protected-milled-polished-sublimed-conductive coated tobacco flower petal.

These procedure now open the way to using the sample exposure mechanisms available from dual beam microscopes on a wide range of temperature-sensitive specimens without causing damage. In addition to the high resolution secondary electron

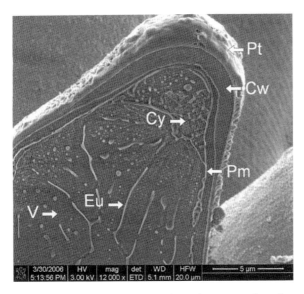

Figure 6.18. SEM image of a frozen-hydrated section of a tobacco flower petal after preliminary sputter coating. FIB-GIS coating-protection, FIB milling, and secondary sputter coating. C = cell wall; Pm = plasma membrane; Cy = cytoplasm; Eu = eutectic ice; V = vacuole; Pt = thin platinum coating layer (Picture courtesy of Mike Hayles, FEI Company, Eindhoven, The Netherlands)

images, the electron beam instrument can also be used to obtain in situ chemical information from the emitted photons and backscattered electrons.

6.4.3 Combined Plasma Etching and Ion Beam Etching

Although it is somewhat invidious in a textbook to isolate just one company as a supplier of a particular piece of equipment, Fischione Instruments in the United States has combined plasma etching and ion beam etching (and sputter coating) into one piece of equipment that completes the final phases of specimen exposure (and cleaning).

Specimens 25 mm in diameter and depth that have first been exposed by mechanical means such as cutting, grinding, cleaving, and initial polishing, are inserted into this novel piece of equipment. Although the equipment was developed for microelectronic samples, there is no reason it cannot be applied to prepare other types of specimens for both SEM and TEM.

Figure 6.19 is of the Fischione Instruments 1030 Automated Sample Preparation System that allows plasma cleaning, ion beam etching, reactive ion beam etching, reactive ion etching, and sputter coating to be carried out in one chamber. This is achieved by giving the sample stage four degrees of freedom to allow the specimen to be moved and tilted to optimize plasma etching and the different ion beam activities of milling, reactive ion etching, reactive ion beam etching, and ion beam sputter coating.

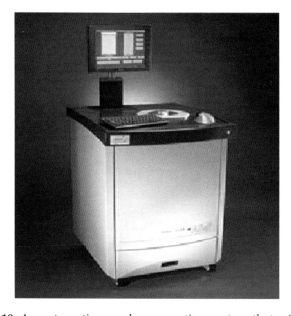

Figure 6.19. An automatic sample preparation system that will plasma clean, ion beam etch, reactive ion beam etch, reactive ion etch, and carry out high resolution ion beam sputter coating (Picture courtesy of Fischione Instruments, Pittsburgh, PA)

The different parameters for all five operations are first preset from a menu that is run automatically once the sample is sealed into the stage. For example, the samples may first be plasma cleaned, scratches and blemishes on the surface are removed by ion beam etching, the now very smooth surface features enhanced by one or more of the reactive ion beam process and finally, the sample may, if necessary, be covered with a very thin conductive layer of metal. The system operates at a very high, clean vacuum, has ports to allow both inert and active gases to be used at the different stages of the sample preparation and accommodate up to four different coating materials. Figures 6.20 and 6.21 show images of a sample before and after treatment.

Figure 6.20. A 2 kV SE image ×30,000 of an oxidized cross section of a copper based electronic material after being mechanically ground and then polished with a 100 nm diamond finish (Picture courtesy of Fischione Instruments, Pittsburgh, PA)

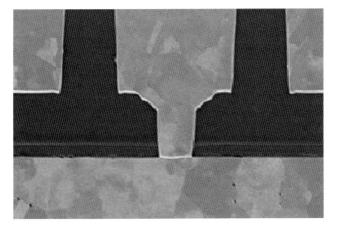

Figure 6.21. A 2 kV SE image at ×30,000 of the same specimen shown in Figure 6.20 following plasma etching, ion beam etching, reactive ion beam etching, and ion beam sputter coated in the Fischione Automatic Sample Preparation system (Picture courtesy of Fischione Instruments, Pittsburgh, PA)

6.5 Suggested Procedures for Exposing the Six Different Types of SEM Specimens

6.5.1 Metals, Alloys, and Metallic Materials

The most usual way to expose these samples is to use the vigorous application of power-driven or hand-held saws followed by a lapping process. The final polishing process may be either purely mechanical or involve chemical intervention, depending on whether the sample is to be imaged and/or analyzed.

6.5.2 Hard, Dry, Inorganic Materials

These types of samples may be treated in the same way as metals, but in a somewhat less assertive manner. The more brittle samples initially may be exposed by chipping, breaking, cleaving, and snapping. The final lapping process should make use of the wide range of mechanical, chemical, and high energy beam etching procedures.

6.5.3 Hard and Firm, Dry Natural Organic Materials

Depending on the specimen's hardness, samples may be exposed either by using hand saws or sharp knives. The lapping processes used in the two previous groups are only applicable to the hardest specimens, and great care has to be taken to remove the abrading grits without wetting the specimen. Chemical and high energy beam procedures should not be used, as they lack little selective sensitivity and will remove the whole specimen. The best way to produce a fine surface is to plane the surface with a sharp hard metal or diamond blade.

6.5.4 Synthetic Organic Polymer Material

The same types of procedures used for hard, dry, organic material can be used with polymer material. Soft polymers are best exposed by hand-held sharp knives followed by planning in a microtome. The conventional lapping procedure for hard specimens should not be used on these types of samples.

6.5.5 Biological Organisms and Materials

These types of samples generally are moist and very soft and easily can be exposed using very sharp knives. The plethora of lapping procedures are not options as they cause more damage than revealing cogent information about the sample. In many cases these types of specimens must first be stabilized and embedded before attempting to expose their fine structural and chemical details.

6.5.6 Wet and Liquid Materials

Depending on how much liquid there is in the wet sample, they must first either be solidified by careful freezing or stabilized and embedded in one of the selected hydrophilic acrylic resins before their internal details may be exposed by cutting with sharp knives. These matters are discussed elsewhere in this book.

7

Sample Dehydration

7.1 Introduction

Strictly speaking, the term "dehydration" is used to describe the removal of water. Each of the six different groups of samples being considered in this book may be damp, wet, or critically associated with water. In addition, some samples may be associated with non-aqueous liquids such as organic fluids, oils, and polymers. Conventional SEM demands that water and organic fluids either must be removed from samples or immobilized before they are examined and analyzed.

It is more usual (and more convenient) to first dry the sample before it is exposed to the arid environment of the microscope column. Drying hard robust materials such as metals and rocks is a relatively straightforward procedure. The surface water may be absorbed by soft paper or fabric tissue, taking care not to disturb the sample surface structure, and then allowing the remaining water to evaporate.

Much more attention has to be paid when drying organic materials, polymers, biological samples and, not surprisingly, wet, moist, and liquid samples. There are several different ways samples may be dehydrated, and much of the discussion that follows applies to removing the ubiquitous compound, water. Before discussing how this may be achieved it is appropriate to briefly consider how wet, moist, liquid samples may be examined directly in the SEM.

7.2 Methods Used to Directly Examine Wet, Moist, Liquid Samples in the SEM

7.2.1 Examination of Wet and Moist Cells in Their Natural Liquid State

The first attempt to look at wet and moist specimens in the SEM was made by Smith in 1956. The subsequent development of this approach has sought to provide, within the microscope column, a discrete chamber maintained as close as possible to the natural environment of the specimen. Thornley (1960) showed that it was possible to image liquid samples in the SEM by sealing the wet material between two thin layers of carbon. This approach has been subsequently developed into so-called wet cell devices, which are small enough to fit onto the microscope stage. Lane (1970) devised a self-contained substage assembly for the SEM that consisted of a chamber with a controlled water vapor leak and a small aperture to admit the incident electron beam and allow the generated secondary electrons to be detected. This approach has developed into the large variable-pressure SEM, sometimes referred to as the environmental SEM.

7.2.1.1 Wet Cells

A number of different wet cells are available for the SEM, and one of the most recent devices is the capsule, developed and marketed by Quantomix™ (www.quantomix.com) based on work done at the Weizmann Institute in Rehovot, Israel.

Figure 7.1. The Quantomix WETSEM® QW-102 for studying wet specimens in a scanning electron microscope (Picture courtesy of Quantomix Ltd., Rehovot, Israel)

7.2 Methods Used to Directly Examine Wet, Moist, Liquid Samples in the SEM

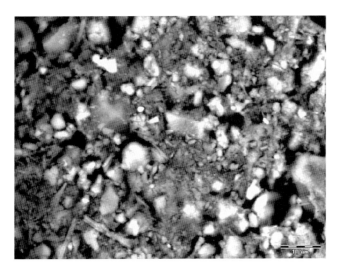

Figure 7.2. A backscattered electron image of Dead Sea mud that has been processed for use as a cosmetic material, examined in the Quantomix WetCell. The wet cosmetic material contains a number of minerals that can be identified by energy dispersive x-ray microanalysis. No preparation is required other than placing the sample in the capsule (Picture courtesy of Quantomix Ltd., Rehovot, Israel)

The capsule is made of a metal-coated plastic, vacuum tight chamber, bounded by a unique, electron-transparent and pressure-resistant membrane. The complete two-part capsule, which is only 13 mm in diameter and 10 mm deep, with an 8-mm basal pin, can be easily fitted onto any standard SEM stage. The 100- to 200-nm membrane (depending on capsule type) completely isolates hydrated samples at atmospheric pressure from the high vacuum in the microscope column, while allowing the primary electron beam to penetrate and the different signals to be collected. Figure 7.1 shows a Quantomix WETSEM® capsule.

To use the WETSEM, the capsule is opened simply by turning the upper sealing stub followed by applying 15 μl of the wet sample onto the surface of the ultra-thin membrane at the base of the lower chamber (the "liquid dish"), and then sealing the capsule by replacing the stub. The capsule is then inverted and the 8-mm pin placed in the hole on the SEM stage. The pressure tight membrane is now uppermost and the layer of liquid sample remains attached to the underside.

Backscattered electron images are the most frequently used imaging mode but secondary electron images and x-ray microanalysis may also be carried out using this device. The capsule has been used to examine liquid samples such as emulsions, foods, oils, paints, inks, and drugs; cosmetic samples such as pastes,

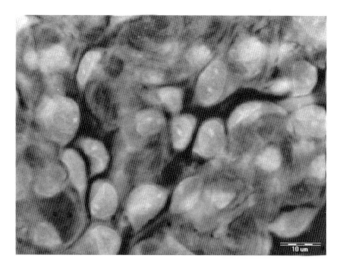

Figure 7.3. Backscattered electron image of rat kidney tissue examined in a Quantomix WETSEM. The specimen was stabilized in formalin, stained in aqueous uranyl acetate, and had not undergone dehydration (Picture courtesy of Quantomix Ltd., Rehovot, Israel)

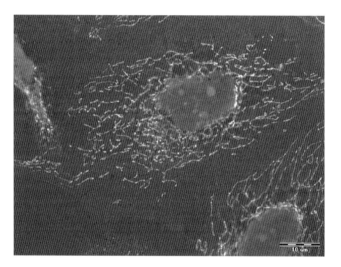

Figure 7.4. A backscattered electron image of HeLa cells, cultured directly in the QX-120 capsule membrane in which the mitochondria have been immunogold labeled. The tissue has not been chemically stabilized or dehydrated after the immunolabeling (Picture courtesy of Quantomix Ltd., Rehovot, Israel) (Goldstein et al. 2004)

powders, and foams; metallic, inorganic, and organic particle in suspension; biological tissues and cells; and a wide range of adherent structures. A suitably prepared sample can be loaded into the capsule and imaged in the SEM within 3–4 min. Figures 7.2–7.4 are images of different types of materials examined in this unique specimen holder.

7.2.1.2 Variable-Pressure Scanning Electron Microscopes

In these instruments, the interior of the chamber is either filled with water vapor or continuously flushed with an inert carrier gas at a pressure range of 10–200 Pa (0.1–20 Torr). A pressure of 20 Torr and 295 K is the saturation pressure of water, and liquid water can be stabilized for observation at this temperature and pressure in the chamber.

Specimens may be examined using secondary and backscattered electrons and analyzed by x-ray microanalysis in this type of microscope because the environment around the specimen no longer has to be at high vacuum. The variable-pressure SEM can be used to study an extraordinarily wide range of specimens, such as watching paint drying, cement setting, and insulating polymers, simply because the specimens need little preparation prior to microscopy. Fully hydrated biological specimens that are either moist or wet may be observed without them drying or shrinking. Suitable samples can be loaded into the microscope and imaged within 6–8 min. A good general up-to-date account of variable-pressure SEM may be found in the book by Goldstein et al. (2004). More detailed information can be found in the up-to-date comprehensive bibliography maintained by Eric Doehne of the Getty Conservation Institute, Los Angeles (edoehne@getty.edu).

7.2.1.3 Infiltration with Low Vapor Liquids

This involves replacing the water in samples by slow infiltration with very low vapor point liquids such as glycerol or triethylene glycol. The only real advantages of this rather messy procedure are that it retains lipids, fats, and waxes, is inexpensive, and is easy to carry out. The paper by Ensikat and Barthlott (1993) details this approach.

7.2.1.4 Low Temperature Microscopy and Analysis

This involves initiating a phase change in wet, moist, liquid samples by converting them to solids that are examined at low temperature on a cold stage in the microscope. An outline of these procedures are discussed in Chapters 8 and 9; full details may be found in Echlin (1992).

7.3 Methods Involving the Removal of Liquids

7.3.1 Air Drying

Water is frequently associated with SEM specimens and drying problems arise because of the high surface tension of water against air. Table 7.1 provides information about water and a number of organic chemicals associated with dehydration.

Table 7.1. Surface tension (Nm^{-2}), boiling point (°K) and flammability of water and some organic liquids

Chemical	Nm^{-2}	°K	Properties
Water	72.75	373	Non-flammable
Acetone	22.70	330	Highly inflammable
Acetonitrile	29.04	355	Poisonous, inflammable
Ethyl alcohol	22.75	352	Inflammable
Methyl alcohol	22.61	338	Poisonous, inflammable
Propylene oxide	22.16	307	Highly inflammable
Liquid carbon dioxide	1.16	304 K (at 5.53 MPa)	Nonflammable

A major problem arises as water is removed from specimens by drying in air. As the water/air interface passes initially from the surface and then through the bulk of the specimen, the surface tension forces associated with the interface can rise to as high as 20–100 Nm (2,000 psi) during the final phases of air drying. These forces become increasingly larger as the structures being dried become smaller. This dramatic increase in surface tension severely distorts the structural integrity of soft specimens in additional to problems with specimens that contain dissolved inorganic and organic solutes. Figure 7.5 shows the type of damage that can be expected in images of air dried specimens:

1. As the water is removed during air drying, dissolved salt solutions become more concentrated and may eventually be precipitated ("salting out"). This can lead to pH changes and, in turn, to changes in the conformation, state of aggregation, and functionality of macromolecules.
2. A related problem arises as water is removed from biological material by air drying. The decrease in the water content causes the osmolarity of cell contents to become gradually hypertonic with a consequent dramatic change in their structure. Rapid drying of wet samples from the aqueous phase by placing them in a vacuum has the added disadvantage of ice crystal formation due to the evaporation of the water at reduced pressure. The formation of ice crystals can distort the specimen. Put briefly, the cells and tissues shrink!

There is a wide range of other techniques for air drying that, although they appear effective for the sample chosen, may not necessarily be widely applicable. For example, Lamb and Ingram (1979) devised a method that allows samples to be directly dried from ethanol under a stream of dry argon. Douchet and Bradley (1989) adopted the mechanisms found in a domestic washing machine. They describe a method by which micrometer size particles, suspended in hexane and placed on a glass cover slip, were spun at 1,000 rpm in a vacuum! Leipins and de Harven (1978) suspended single cell samples in the fluorocarbon Freon 113 following dehydrated in ethanol and evaporated the liquid at a

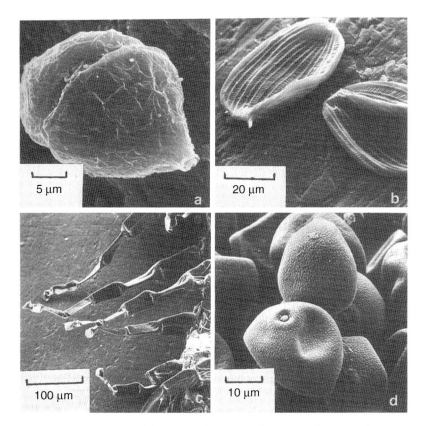

Figure 7.5. Examples of the gross distortion due to air drying soft specimens. A. Dinoflagellate alga. B. Euglenoid alga. C. Leaf hair of deadly nightshade. D. Grass pollen grains

pressure of 3 Pa (3×10^{-2} Torr). Similar experiments were carried out by Nation (1983) using hexamethyldisilazane. This farrago of methods is included here, not as a binding recommendation, but to illustrate the variety of approaches to air drying samples.

There is little to be gained by heating wet organic and biological specimens during drying, for even at 100°C the surface tension of water against air is only reduced to 58.9 Pa. Air drying samples from some of the volatile organic liquids mentioned above which have an appreciably lower surface tension than water, diminish, but do not negate, the problems associated with surface tension. In addition, these organic materials are inflammable and must be kept at room temperature and only allowed to dry in a properly vented flameproof fume cupboard. The one exception is liquid carbon dioxide that, although it has a very low surface tension, requires special instrumentation that will be discussed later.

In summary, air drying should only be used on wet samples that are sufficiently robust to withstand the deleterious effects of the high surface tension of water.

7.3.2 Chemical Dehydration

There are two chemical ways to removed water from specimens. It may be dissolved in organic chemicals or reacted with organic chemical to form a non-aqueous organic liquid.

7.3.2.1 Passive Chemical Ways of Dissolving Water

There are a wide number of different protocols; two general schedules are illustrated.

7.3.2.1.1 Generic Dehydration of Samples to Be Directly Examined in the SEM
1. Wash small (1- to 2-mm^3) stabilized samples several times in distilled water to remove any stabilization liquids. Place them in a small sealed glass container containing distilled water, approximately 50 times the volume of the specimens.
2. Gradually add a sufficient volume of 100% ethyl alcohol to ensure a 30% ethanol solution. Replace the container seal and leave for 10–15 min at room temperature.
3. Using a pipette, rapidly remove nearly all the 30% solution and immediately replace with a same volume of 70% ethanol. The specimens should not be allowed to dry. Replace the container seal and leave for 10–15 min at room temperature.
4. Repeat the process with 95%, 100%, and twice with anhydrous ethanol. The firmly sealed samples may remain briefly at room temperature before proceeding to the next stage in the process.

The anhydrous ethanol is very hygroscopic and should be kept over a suitable drying agent, and care taken not to shake this solution when dispensing the very dry alcohol as this may cause fine particles of the drying agent such as Drierite to be transferred to the specimen. The temperature of dehydration does not appear critical, although it is usual to carry out the procedure at the same temperature used for sample stabilization.

This generic dehydration procedure has a wide number of variants. Methyl alcohol, acetone, or acetonitrile (Edwards et al., 1992) may be used instead of ethyl alcohol. The samples may be gently rotated during the procedure. The time may be varied depending on the size, softness, and permeability of the specimen. The dehydration process may either proceed through a series of discrete steps as shown above or by a continuous and gradual change in the concentration of the dehydrating chemical.

7.3.2.1.2 Rapid Dehydration of Samples to Be Directly Examined in the SEM (Bencosme and Tsutsumi, 1970)
1. Wash very small (0.1 mm^3) stabilized samples in distilled water to remove any stabilization liquids and place them in a small sealed glass container containing distilled water, approximately 50 times the volume of the specimens.

2. Using a pipette, rapidly remove nearly all the distilled water and immediately replace with 70% ethanol. Replace the container seal and gently agitate for 3 min at room temperature.
3. Repeat the process with 80% and 90% ethanol followed by two 5 min changes in 100% ethanol. The firmly sealed sample may remain briefly at room temperature before proceeding to the next stages.

Samples dehydrated by these two approaches finish up suspended in a relatively volatile organic liquid and the drying may be completed in two different ways:

1. The dehydrated specimens may either be substituted into a more highly volatile liquid such as diethyl ether, dimethyl ether, or 1–2 epoxy propane and then carefully air dried as described in the previous section. Alternatively, the dehydrated specimens may be substituted into tetra-butyl alcohol, which is then removed under low vacuum. The same procedure works with tetramethylsilane. Both methods are rapid, but some shrinkage must be expected. These rapid drying approaches are not recommended for soft specimens.
2. The drying may be completed either by critical point dryer (see the next section) or by slowly infiltrating the sample in a liquid resin, which can be subsequently polymerized as described in the following.

7.3.2.1.3 Dehydration of Samples to Be Embedded in Resin That Is Polymerized and Subsequently Fractured or Sectioned Before Being Examined in the SEM

The process of embedding material is discussed in Chapter 5. A wide range of polymers are used as embedding agents. Some are miscible with water, others with ethanol, acetone, propylene oxide, and acetonitrile. Many of the more popular, and very effective, epoxy resins are not entirely miscible with the common dehydrating agents and it is usually necessary to interpose an intermediate agent that is miscible with both the dehydrating chemicals and the liquid resin. The most useful intermediate fluid is acetonitrile. Propylene oxide, which is miscible with ethanol and epoxy resin, also can be used but is very inflammable and carcinogenic.

Properly dehydrated samples are very dry and it is important not to expose them to the atmosphere as they will rapidly absorb water vapor. This presents a potential problem when transferring the dry specimen suspended in the very dry dehydration liquid to other instruments involved in the specimen preparation process and, eventually, to the SEM. The pragmatic approach it to bring the dried sample as close as possible to the instrument, i.e., a coating device and carefully and quickly pick up the sample, remove the *excess* of the dehydration fluid, place it on the specimen holding device, and quickly evacuate to a low pressure. This final process of drying, which is now from an organic liquid rather than water, is less likely to cause any further structural or chemical changes, although the potential for shrinkage remains.

7.3.2.2 Reactive Chemical Ways of Removing Water

Muller and Jacks (1975) devised a method for rapid dehydration based on the exothermic chemical reaction of water with acidulated 2,2-dimethoxypropane (DMP) to form acetone and methanol. Table 7.2 shows how the procedure may be carried out.

The only advantage of this approach is speed, because the amount of shrinkage in soft samples is more than in other chemical dehydration methods and there is a greater extraction of lipids. This procedure should only be used when speed is important and detailed structural integrity is not an issue. Figure 7.6 shows images of soft tissue that have been slowly dehydrated in ethanol and rapidly dehydrated in di-methoxypropane prior to critical point drying (CPD).

Chemical dehydration involves the use of different chemicals in many different ways on many different samples. Such a wide variation inevitably leads to some structural damage and loss of chemical integrity in nearly all but the most robust samples. These changes are most apparent in biological, moist, and wet specimens because water plays such a pivotal role in their structure and functionality, and its removal can cause drastic changes. In addition, the organic dehydrating agents cause chemical changes and losses in many of the lipids, carbohydrates, and proteins even in stabilized living samples. Yet, in spite of these short comings, chemical dehydration, although far from perfect, is an effective way of removing water from specimens.

7.3.3 Critical Point Drying

This technique, strictly speaking, is not a dehydration process but is designed to help overcome the problems associated with

Table 7.2. A procedure for a rapid chemical removal of water from a specimen

Rapid dehydration
1. The specimen must first be thoroughly washed in distilled water to remove any traces of phosphate buffers that will precipitate in DMP (dimethyl-sulfoxide)
2. 0.05 ml of concentrated hydrochloric acid is added to 100 ml of DMP
3. Small samples of materials are added to 1–5 ml of the DNP solution, which is cooled to 278 K
4. The specimens are left in the solution for 15 min to 16 h at room temperature depending on their size
5. The dehydrated sample is transferred to anhydrous acetone or methanol and kept in sealed container

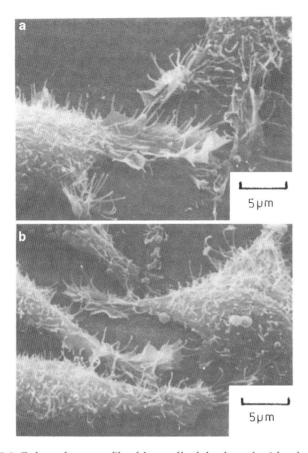

Figure 7.6. Cultured mouse fibroblast cells dehydrated with ethanol *(A)* and with 2-2-dimethoxypropane *(B)* (Goldstein, et al.)

surface tension effects when specimens are dehydrated using some of the organic liquids mentioned in the previous section. These liquids are excellent dehydrating agents, but for specimens to be directly examined in the SEM, it is difficult to transfer the bone dry sample, suspended in the anhydrous liquid, through air into the microscope or ancillary equipment without the problems associated with surface tension. Critical point drying overcomes these problems. Critical point drying was introduced to transmission electron microscopy more than half a century ago by Anderson (1951) and was adopted for the SEM in the late 1960s.

Critical point drying is based on the principle that there is a critical point on the isothermal for many liquids at which the density of the liquid and gaseous phases is identical. At this critical temperature-pressure point there is an equal exchange of molecules between the gas and liquid phases and surface tension is zero. Table 7.3 gives data for some of the liquids involved in critical point drying.

Table 7.3. The critical temperature and pressure of chemicals which may be used in critical point drying

Chemical	Critical temperature (K)	Critical pressure (MPa)
Water	648	22.12
Methanol	513	7.95
Freon 13	302	3.87
Carbon dioxide	304	7.39

A number of different units are used to express pressure. The numerical data in this book use SI metric units, the internationally accepted units in science. Unfortunately, the literature still contains some of the older units and these may be converted to S.I. units as follows:

- To convert standard atmosphere to Mpa, multiply by 1.0133^{-1}
- To convert pounds per square inch to MPa, multiply by 6,895

The high critical temperature for water prevents it being used directly with soft and biological material even though its critical pressure is low. The critical temperature and pressure for methanol and other similar liquids used for dehydration is also too high. In contrast, the critical temperatures of both Freon 13 (a fluorocarbon refrigerant) and liquid carbon dioxide are low and although the critical pressure for Freon 13 is low, the pressure for carbon dioxide is high.

From this information it would appear that Freon 13 and other similar fluorocarbons would be the best to use for critical point drying. However, the images in Figure 7.7 show that better structural integrity is retained when liquid CO_2 is used rather that fluorocarbons. The fluorocarbons are banned as refrigerants because they are directly related to the destruction of the ozone layer and consequent global warming and are no longer generally available. For this reason, liquid carbon dioxide is used for critical point drying.

Liquid carbon dioxide is miscible with 100% ethanol, methanol, and acetone, and samples totally immersed in one of these chemicals after dehydration can be transferred direct to liquid carbon dioxide. There is no need to use an intermediate fluid, such as amyl acetate or one of the now banned fluorocarbons, between the organic dehydration liquid and carbon dioxide. The only possible advantage of using amyl acetate is that it has a characteristic "pear drops" smell, which can be used as a guide that the dehydrating alcohols are finally removed during critical point drying.

The critical point apparatus is complicated and the procedures are potentially hazardous because high liquid and gas pressures are involved. A high pressure vessel is water cooled and equipped with a sturdy port to insert the sample, inlet and outlet

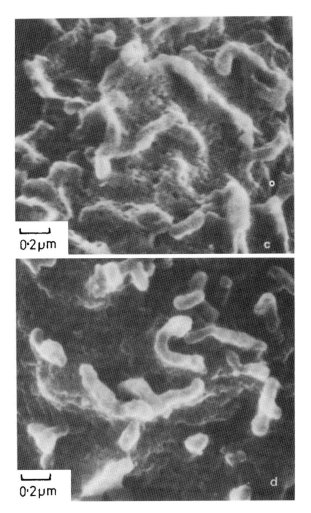

Figure 7.7. Comparison between the surface of a mastocytoma cell that had been critical point dried using CO_2 (C) and critical point dried using a fluorocarbon (D). The CO_2 critical point dried sample shows minute pores in the cell surface that are absent in the fluorocarbon treated sample (Picture from Goldstein et al., 1992)

valves for liquid and gaseous CO_2, a small heater, thermocouple, and a pressure gauge. This high pressure vessel is linked to a liquid CO_2 cylinder fitted with a siphon tube to ensure that liquid CO_2 is fed into the critical point dryer.

There are a large number of companies who manufacture and market critical point dryers, including Emitech Ltd. (www.emitech.co.uk), Structural Probe Inc. (www.2spi.com), Tousimis MEMS Products (www.tousimis.com), Ladd Research (www.laddresearch.com), and ProSciTech (www.proscitech.com.au). Figure 7.8 is of a critical point dryer made and marketed by Quorum Technologies (www.quorumtech.com). Figure 7.9 is a diagram of the components of a critical point dryer.

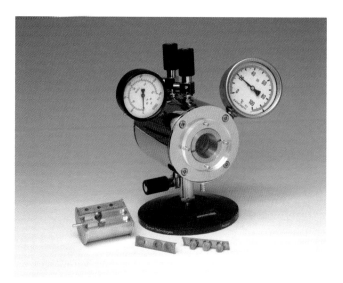

Figure 7.8. The Quorum Technologies E3100 large chamber critical point dryer with specimen containers and the boat to transfer the specimens into the critical point drier (Picture courtesy of Quorum Technologies, Ringmer, Sussex, UK)

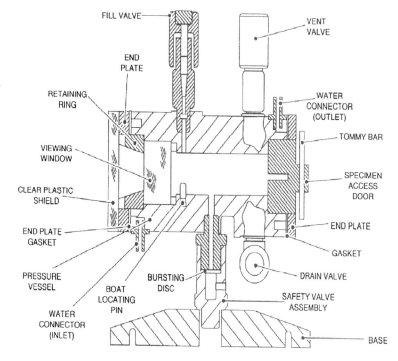

Figure 7.9. Sectional diagram of a critical point dryer (Picture from Goldstein et al., 2004)

The equipment, sometimes referred as a "bomb," is potentially very hazardous because it works at such high pressures. It should only be used by trained personnel who should follow the maker's instructions. The following generic instructions will help the reader to understand the sequence of actions during CPD and where particular caution should be taken:

1. Carefully read and understand the maker's instructions. Make sure all the valves and other connection are secure. Make sure the O-rings on the door to the CPD are undamaged and fully functional. Make sure that the cylinder of liquid carbon dioxide is fitted with a siphon tube with an opening at the bottom of the inside of the cylinder to ensure delivery of liquid carbon dioxide.
2. Make sure that the inside of the CPD is dry and at room temperature (298 K). Check that it is connected to a cylinder of liquid carbon dioxide (CO_2); to the cooling tap water; the electrical power and that all the valves and taps are closed.
3. The dehydrated specimens that should be immersed in anhydrous ethanol (or methanol or acetone) are quickly transferred to the CPD sample holder, which is also filled with dry ethanol. The loaded specimen holder quickly placed into the CPD chamber and the door firmly and securely closed.
4. Cool the now loaded and sealed CPD chamber to 293 K. Open the valve to the cylinder of liquid CO_2 and ensure that the pressure is 5.53 MPa (800 psi). Slowly open the *inlet valve* to the CPD chamber and fill it nearly to the top with liquid CO_2. The chamber pressure will rise to 5.53 MPa and the liquid–gas meniscus of the ethanol–carbon dioxide mixture will rise inside the chamber. This process may be observed in systems fitted with viewing port. The *inlet valve* should remain open.
5. Allow the mixture to stand for a few minutes and then gently open the *vent valve* to the chamber to allow fresh liquid CO_2 to flush through the sample chamber. It is most important that the samples remain below the meniscus to prevent exposure to the gas phase of the CO_2. Continue intermittent flushing for 10–15 min to ensure the ethanol in the specimen is fully substituted with liquid CO_2.
6. Close both the *inlet and vent valves* and leave the sample to fully infiltrate with the liquid CO_2. This will take about an hour with a 1-mm³ sample, but the time will need adjusting depending on the size, density, and permeability of the specimen. During the infiltration stage periodically check that the sample remains below the surface of the liquid CO_2.
7. Re-open the *inlet valve* and fill the chamber with a final flush of fresh liquid CO_2 and slowly adjust the *vent valve* to ensure the chamber fills to about 75% of its capacity. Close both the *inlet and outlet valves* to the chamber and the valve on the CO_2 cylinder.

8. Turn off the cooling water and slowly heat the chamber to 308 K. The pressure will rise and the CO_2 will pass through its critical point of 7.39 MPa and eventually reach a pressure of 8.27 MPa (1,200 psi). This process may be observed through the chamber window and the liquid meniscus disappears as the CO_2 passes through the critical point.

9. Maintain the chamber temperature and *very slowly* open the *vent valve* so that the now gaseous CO_2 from the chamber is released. If the venting is too fast, the gas phase will momentarily revert to the liquid phase, which will wet and damage the specimen.

10. Once the chamber has reached atmospheric pressure, quickly remove the now very dry specimens and immediately transfer them to a sealed container placed in a desiccator. This transfer should be done as quickly as possible because the now anhydrous sample can quickly absorb atmospheric water.

11. Ensure the door to the chamber is slightly open, and then slowly open the CO_2 *inlet valve* to the chamber to allow the residual CO_2 in the line from the CO_2 cylinder to be vented. Turn off the heater and once the chamber has reached room temperature, replace the chamber door to ensure the inside of the chamber remains dry.

7.3.4 Disadvantages of Critical Point Drying

Although critical point drying overcomes the problem associated with the deleterious effect of surface tension, there is compelling evidence that liquid dehydration followed by critical point drying can damage many soft specimens in two different ways.

7.3.4.1 Structural Integrity

A series of detailed studies (Boyde et al., 1977; Boyde and Maconnachie, 1979) show that critical point drying causes gross (up to 70%) and spatially unequal dimensional changes in most specimens. Boyde and Maconnachie (1979), Woolweber et al. (1981), and Gamliel (1985a) discuss the advantages of adding various heavy metals, high density elements, and organic additives, which may be included in the stabilization procedures and can reduce shrinkage to 5%. However, the selective erosion and perturbation of surface features still remains a problem.

7.3.4.2 Chemical Integrity

A quick search of the appropriate literature will reveal an extensive chemical engineering literature that shows that liquid CO_2, usually referred to as supercritical CO_2, is a powerful solvent. For example, it can be used to degrease metals; dissolve metal ions, lipids, proteins, and carbohydrates; decaffeinate coffee and tea; and even remove nearly all the nicotine from cured tobacco leaves.

It would be wrong to suggest that CPD always give rise to distorted samples, although this is not a method that should be used for high resolution structural studies on delicate organic specimens or with analytical investigation on any type of specimen. At low magnification, i.e., up to 3–5,000×, critical point drying is probably satisfactory for structural studies. Figure 7.10 shows that critical point drying is a much better procedure than simple air-drying as a dehydration procedure.

7.3.5 Low Temperature Drying

This process is one of several different low temperatures procedures that may be used to prepare specimens for electron microscopy. A brief explanation of these different methods is necessary before providing details of freeze-drying:

1. *Freeze-drying*, is a *physical* dehydration method by which frozen water (ice) is removed by low temperature sublimation in a high vacuum.

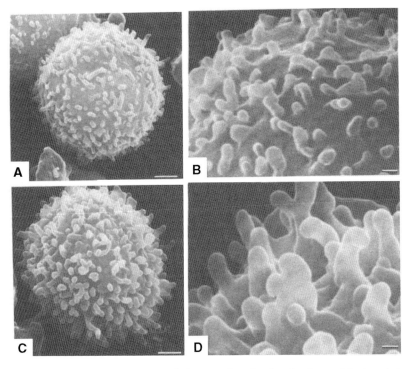

Figure 7.10. A comparison of the results obtained after critical point drying *(A and B)* and freeze-drying *(C and D)* of lymphocytes showing narrow microvilli and a relatively smooth underlying surface. The microvilli are much wider and the cell surface much smoother in the freeze-dried specimens. Magnification bar = 200 nm (Picture from Goldstein et al., 1992)

2. *Freeze-substitution* is a *chemical* dehydration process in which ice in a frozen hydrated specimen is dissolved in an organic solvent at low temperatures. This is discussed in the next section.

7.3.6 Freeze-Drying

How freeze-drying is used to dehydrate and preserve food material and pharmaceutical agents is not discussed here as it centers on slow cooling; the recent book by Franks (2007) is an excellent introduction to the physiochemistry of freeze-drying associated with these processes. We consider the way freeze-drying is used to preserve fine structural and chemical identity in specimens that depend on fast cooling. The physiochemistry of these matters is discussed in the books by Robards and Sleytr (1985) and Echlin (1992).

7.3.6.1 The Consequences of Freeze-Drying

Freeze-drying for microscopy and analysis will probably always remain an empirical process, although general guidelines may be followed. The experimental protocols have to be designed *ab initio* around the sample concerned, the expectations of the image quality, and the veracity of the chemical analysis. There is more to freeze-drying than just removing the water from a specimen by sublimation.

Most molecules in hydrated systems are surrounded by a layer of water, referred to as the *hydration shell*, which prevents the molecules from sticking together. This is a particularly important phenomenon in biological material and some organic materials, such as emulsions and adhesives, which have an affinity for water. The removal of the hydration shell may lead to aggregation and/or collapse of the macromolecular organic matrix. These changes are a consequence of thermal movements and the natural stickiness of dehydrated biological material and some organic materials.

In addition to the water associated with the hydration shell, there is the *bulk* or largely *unperturbed water*, which makes up a substantial part of the water in many hydrated systems. Removal of the bulk water is related to shrinkage, which can be a serious problem in freeze-drying, for unlike chemical dehydration, no attempt is made to replace the water with another material. The relative seriousness of shrinkage or collapse is, of course, related to the resolution of the imaging system. At the level of the SEM and analysis the consequences of the molecular changes are all too evident.

Nearly all freeze-drying protocols are two-step processes in which the temperature of the sample is eventually raised to ambient, or above, at the end of the drying process to remove the residual moisture.

Primary drying centers on the direct sublimation of ice formed as a consequence of freezing the bulk water in the sample. About 90% of the initial water is lost during this phase, leaving behind a highly porous "ghost" of the initial hydrated structure.

Secondary drying, sometimes referred to as "isothermal desorption," removes the remaining water associated with macromolecules. This is a tedious operation that may take as long as the primary drying phase, as water is desorbed from the highly porous macromolecular matrix.

Freeze-drying is carried out in situations in which the solidified water is in a crystalline state. Chapter 8 shows that very fast cooling rates of very small specimens will convert liquid water to vitreous ice. There is some evidence (see Echlin 1992 for details) that it might be possible to sublimate vitreous ice at very low temperatures and pressures over a very long period of time.

Nearly all the calculations and predictions for freeze-drying have been made for pure water, a liquid that rarely exists in nature because water is a near universal solvent. The presence of solutes and embedded macromolecules raise the temperature range over which crystalline transitions are likely to occur and will slow down the rate at which water is sublimed from the frozen state. In addition, it is very difficult to measure the temperature at the surface from which the ice is subliming, and although there may be an accurate measurement of the specimen holder, the subliming ice surface is probably significantly warmer. A useful rule of thumb is to assume there is a 10° temperature difference at each metal–metal contact in a cooling device. For example, if the cold stage is at 130 K, the metal specimen holder will be at 140 K and the specimen itself will be at 150 K.

7.3.6.2 The Theoretical Basis of Freeze-Drying

A lot has been written about the theory of freeze-drying and a brief discussion is included here to help explain the large deviation between what theory predicts and what actually occurs before discussing the practical approach to freeze-drying specimens for microscopy and analysis:

1. All specimens have a finite thickness and the water molecules have to be removed through the fine structure and porous network of the specimen. As the drying proceeds, the layer of dried material becomes progressively thicker and the pathway through which the escaping water molecules have to pass becomes progressively longer. It has been calculated that it would take water molecules diffusing through a 1 mm^3 of frozen liver 1,000 times longer than diffusing through a 1 mm^3 of crystalline ice. Different samples dry at different rates. The difference between the drying time of a sample compared with the drying time of a similar size and shape

of a piece of crystalline ice is known as the *prolongation factor*. There is no standard rate of diffusion, and it is difficult to calculate just how long it takes to freeze-dry a sample.

2. The sublimation of water needs energy, which must be transferred to the drying boundary. The drying front can proceed through the sample in two different ways:

    a. If the source of heat is radiant energy from above, the heat transfer will take place through the progressively thicker dried layer of material on top of the underlying frozen-hydrated sample. The rate of drying becomes progressively slower the deeper the drying front proceeds into the sample.
    b. If the source of heat is from below and through the sample, the rate of drying will be primarily influenced by the mass transfer of water out of the sample. In practice, all combinations of heat transfer and mass transfer occur and the drying rate depends on:

        i. The size and structure of the sample
        ii. The thermal contact the sample has with the source of heat
        iii. The vacuum condition inside the freeze-dryer
        iv. The efficiency of the system used to trap the sublimed water molecules

   Table 7.4 shows the wide variation of freeze-drying as a function of temperature and pressure.

3. Water associated with the hydration shell surrounding macromolecules does not freeze and can only be effectively desorbed by raising the temperature. This may cause collapse and aggregation in soft samples, and this secondary drying should be only occur over a long period of time.

Table 7.4. Time taken to sublime ice at different temperatures and pressures

Temperature (K)	Vapor pressure	Etching rate (Nm s^{-1})	Drying time (s mm^{-1})	Time to remove 1 mm of ice
213	1.10 Pa	1.48×10^{-3}	6.76×10^{-1}	0.68 s
203	259 mPa	3.64×10^{-2}	2.75	3.0 s
193	53.6 mPa	7.70×10^{-1}	1.30×10^{1}	13 s
183	9.32 mPa	1.37×10^{-1}	7.82×10^{1}	1 min 13 s
173	1.32 mPa	2.00	4.99×10^{-2}	8 min 19 s
163	457 mPa	2.30×10^{-1}	4.34×10^{-3}	23 min 14 s
153	12.4 mPa	1.99×10^{-2}	5.02×10^{-5}	14 h
143	73.9 nPa	1.22×10^{-3}	8.17×10^{-5}	10 days
133	28.8 nPa	4.95×10^{-3}	2.02×10^{-7}	31 weeks
123	668 pPa	1.2×10^{-4}	8.40×10^{-8}	27 years
113	8 pPa	1.5×10^{-5}	6.00×10^{-10}	2 centuries

From Umrath, 1983.

4. The withdrawal of the latent heat of evaporation causes the sample to cool during freeze-drying with a concomitant decrease in the ice sublimation rate. As the temperature of the sample is raised, the temperature of the drying sample first *decreases* as the water sublimes and then slowly rises to reach a higher equilibrium temperature. These subtle temperature changes may cause the subliming water molecules to condense and wet the sample and undo all the attempts to preserve the sample structural and chemical integrity.

This information provides the basis for designing freeze-drying protocols.

7.3.6.3 Practical Freeze-Drying Guidelines

1. The specimens should be as small as practicable and quench cooled as fast as possible to ensure that, at least, the outer 15- to 20-μm region of the sample is likely to contain either vitreous or microcrystalline ice.
2. Ideally, the sample should be freeze-dried at the lowest possible temperature to diminish re-crystallization.
3. The actual temperatures that are used are a balance between:
   a. The lower the temperature, the longer it takes to dry the specimen.
   b. The lowers the temperature, the lower the chance of re-crystallization damage.
4. The better the efficiency of the water vapor trapping device, the faster the sample is freeze-dried.
5. Freeze-drying should be a slow, continuous process.

7.3.6.4 A Freeze-Drying Protocol

An examination of Table 7.4 will show that at 173 K (−100°C), a 10-μm layer of ice will sublime in about 30 min, whereas at 213 K (−60°C) the same amount of ice will sublime in about 10 s. For a 1-mm^3 cube of biological materials or a soft porous organic material with a prolongation factor of 50, the drying time would be as follows:

1. At a temperature of 173 K and a pressure of 400 mPa (3×10^{-3} Torr) the primary drying would take about 3 days to remove the water from the outermost 30-μm layer of the cube of frozen material.
2. If the temperature is now raised to 213 K, the secondary drying of the remaining 850-μm^3 cube of frozen material would be freeze-dried in just under 5 h assuming that the water sublimes from five faces of the cube of material.
3. If the sample is particularly fragile and susceptible to shrinkage and collapse during the second phase of drying, the

temperature should be lowered to 203 K and the remaining bulk of frozen material should be dried in about 30 h.
4. Once the primary and secondary drying are completed, the freeze-dryer is allowed to slowly rise to ambient temperature, while still maintaining the high vacuum.

This protocol assumes that all the sublimed water is trapped and the sample temperature is close to the cold stage temperature. This protocol should also only be treated as a staring point, for there many variables must be considered; for example, the size, porosity, prolongation factor, water content, and number of samples; the final destination of the sample once it is dried; and whether the drying process is for imaging or analyzing the sample (or both).

7.3.6.5 Equipment for Freeze-Drying

Many freeze dryers are available. The following companies make freeze dryers specifically for preparing specimen for microscopy and analysis: Quorum Technologies (www.quorumtech.com), Emitech Ltd. (www.emitech.co.uk), and Bal-Tech (www.bal-tech.com), whose picture is shown in Figure 7.11.

Figure 7.12 is a schematic image of the essential features of a freeze dryer suitable for preparing small sample for examination in different types of microscope.

The freeze-dryers described in the preceding are all cooled by liquid nitrogen, and in principle can be used with any type of specimen that contains water. In many non-biological specimens, ice crystal damage is less of a problem because the non-aqueous parts of the sample are sufficiently robust to resist the structural damage that water crystallization may cause. Mud, wet wood, and fabric are examples of these types of samples from which water can be removed by freeze-drying.

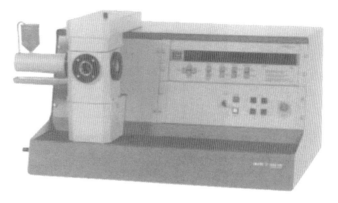

Figure 7.11. Bal-Tech MED 020 Modular Freeze-dryer Unit (Picture courtesy of Bal-Tech AG, Balzers, Lichtenstein)

Such robust wet specimens can be freeze-dried in a Peltier cooled freeze-dryer, which is simpler than a liquid nitrogen freeze-dryer. Figure 7.13 is the K750X freeze-dryer with a Peltier Cooling and Warming Stage from Quorum Technologies

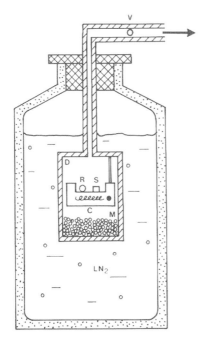

Figure 7.12. Diagram of a liquid nitrogen cooled freeze-drying container. V = vent, D = drying chamber. R = frozen embedding medium, S = frozen sample, C = specimen container, M = molecular sieve, LN_2 = liquid nitrogen (Picture from Echlin, 1992)

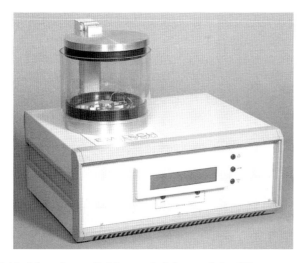

Figure 7.13. A bench-top Peltier cooled freeze drier (Pictures courtesy of Quorum Technologies, Ringmer, Sussex, UK)

(www.quorumtech.com). A Peltier cooler is based on the phenomenon that there is a drop in temperature when a current flows between two different metals or semiconductors. Reversing the current flow causes heating. Peltier coolers have no moving parts and can operate between 223 and 323 K at ambient and low pressure.

7.3.6.6 A Liquid Nitrogen Cooled Freeze Drying Procedure

1. The drying chamber should have a *conductive metal platform*, i.e., copper cooled by liquid nitrogen to at least 173 K. The cold platform should be fitted with a heater, an accurate temperature controller, and a thermostat. Facilities should be available to back fill the chamber with dry nitrogen gas.
2. The *vacuum system* should be capable of maintaining a pressure of 1–10 mPa (5×10^{-6} to 1×10^{-4} Torr). A turbo-molecular pump backed by a rotary pump provides the best vacuum. The vacuum lines should contain either a chemical, i.e., activated alumina, or a low temperature system to trap both water vapor and back-streaming hydrocarbon vapors.
3. The drying chamber must contain an efficient *trapping device* to solidify and trap the water vapor that sublimes from the sample during freeze-drying. As a rule of thumb, the saturation vapor pressure of ice on the trapping condenser should be three orders of magnitude lower than the saturation vapor pressure of ice at the drying temperature. The easiest and most efficient trapping device is a copper plate, cooled with liquid nitrogen (77 K) 5 mm from the drying specimen. At this temperature, the partial pressure of water vapor pressure is 1 aPa (10^{-18}) or 1×10^{-21} Torr. In comparison, the partial pressure of the specimen drying at 173 K is 4 mPa (10^{-5} Torr).
4. In addition to the three important components necessary for drying, it is necessary to have facilities to maintain the *extremely hygroscopic* nature of the freeze-dried material in a very dry environment. If the dried sample is to be embedded in resin, facilities should be available inside the freeze-dryer to enable the initial phases to be carried out in a dry environment. If the dry sample is to be examined in the SEM, it is useful to have an air lock system to enable the sample to be transferred to a suitable coating device and then to the SEM without exposure to air.

In addition to the four essential parts of a freeze-dryer there is one additional, but optional, component that can be useful when following the drying process. A small mass spectrometer attached to the drying chamber can be used to measure the changes in the partial pressure of water vapor as the drying process proceeds. The output of the mass spectrometer can be linked to a computer and a feedback loop can be used to actually control the warming rate of the sample process during freeze-drying.

In addition to the freeze-dryer described here, which operates at 173–213 K and 400 mPa (3×10^{-3} Torr), there are three other types of dryer. The high vacuum (1 mPa or 8×10^{-6} Torr); the low vacuum (1 Pa or 8×10^{-3} Torr), and the cold molecular sieve (molecular distillation; 7 nPa or 10^{-11} Torr). Details of these additional freeze-dryers are described in Echlin (1992).

Freeze-dryers that are cooled by a Peltier device are probably less useful for preparing specimen for high resolution microscopy because the lowest temperature reached by these devices is 213 K. At this temperature, any amorphous ice that had been formed during the initial specimen quench cooling would have very quickly been converted to quite large hexagonal ice crystals, which would distort and damage the sample. If, however, high resolution images are not the prime reason for freeze-drying, then the drying rate of crystalline ice is on the order of 2.5 mm/h at a pressure of 400 mPa (3×10^{-3} Torr).

7.3.6.7 Freeze-Drying from Non-aqueous Solvents

So far this discussion has only considered the freeze-drying process in which the natural water content of a specimen is first transformed to ice, which is then removed by sublimation at low temperatures and high vacuum. In the earlier section on chemical dehydration, it was shown that water may be dissolved and replaced by organic liquids such as ethanol. The anhydrous sample can then be processed by critical point drying, air drying, or infiltrated in a liquid resin that is subsequently polymerized. One additional route may be followed to continue and complete the dehydration process.

The liquids in samples that have initially been slowly chemically dehydrated in ethanol, acetone, or methanol are gradually replaced with other organic liquids, which in turn are removed by a process of *low temperature solvent drying*. The samples are cooled in liquid nitrogen. Fast cooling is not necessary, for there is no water to form ice crystals. Depending on the material, the cooled organic liquid will form either a glass or a microcrystalline solid. The samples may be dried by sublimation at temperatures and pressures that depend on the nature of the material. Table 7.5 shows that most of the organic liquids that are used have a sublimation point very close to their melting point; and in order to have a margin of safety, the sublimation should be carried out 10 K below the melting point. Freeze-drying from non-aqueous solvents is not generally applied to biological specimens.

The low temperature solvent-drying may be carried out in either a liquid nitrogen cooled or a Peltier cooled freeze-dryer or at room temperatures in a modified vacuum evaporator. A liquid nitrogen cooled condenser should be used to trap the sublimed solvents. Condensed solvents that are either toxic or flammable must be vented in a flameproof fume cupboard.

Table 7.5. Properties of materials used for solvent drying by sublimation

Material	Melting point (K)	Sublimation temperature (K)	Sublimation pressure
Acetonitrile	227	226	130 Pa (1 Torr)
Acetone	177	±173	1 Pa–1 mPa (10–3 Torr)
Ethanol	159	±173	1 Pa–1 mPa (10–3 Torr)
Diethyl ether	157	±173	1 Pa–1 mPa (10–3 Torr)
Propylene oxide	161	±173	1 Pa–1 mPa (10–3 Torr)
Amyl acetate	202	192	630 mPa (10–3 Torr)
Camphene	318 (sublimes)	298	130 Pa (1 Torr)

7.3.6.8 Damage and Artifacts Associated with Freeze-Drying

Although freeze-drying is generally considered to be the gentlest way to remove water from biological, moist, and wet samples, specimen damage and artifacts can occur. The damage usually happens during the various phases prior to the actual drying process. Faulty or inadequate chemical stabilization, failure to remove soluble and particulate surface components, and poor sample quench cooling create their own catalog of problems. The damage and artifacts resulting from freeze-drying can be considered at three levels.

7.3.6.8.1 Molecular Artifacts

Collapse and aggregation are exacerbated by the strong electrostatic interactions that may develop during freeze-drying hydrated samples and their support by thermal agitation of macromolecules freed of their hydrophilic exclusion zones. As the water is sublimed, the electrostatic forces between charged groups increase by a factor of 80 owing to the drop in the dielectric constant of water. Franks (1985) stresses the importance of maintaining the hydration shell of proteins during conditions of lower water activity. These problems, which are of concern in the food and pharmaceutical industry, can be partially ameliorated but are largely ignored in non-biological organic material. One way around this problem is to increase the solute concentration, but this may well be incompatible with some biological samples.

7.3.6.8.2 Structural Artifacts

Because water is such an important structural component of hydrated samples, its removal can lead to mechanical instability. As the sublimation front proceeds through the sample, cracking, collapse, shrinkage, wrinkling, and folding can occur. There is a large variation in the structural artifacts that is primarily associated with the secondary drying phase and probably related to the inherent lack of mechanical strength of the material involved and their water content.

Artifacts caused by crystallization during the initial cooling, ice crystal growth during faulty low temperature storage, and accidental re-crystallization during drying usually appear as voids in the dried sample matrix. It is difficult to establish an accurate cause and effect. The crystallization–sublimation–re-crystallization phenomenon is a struggle between the two competing thermodynamic processes. Crystallization and re-crystallization reduce the entropy of the system, sublimation increases it, and in a given sample it is simply a race to see which set of conditions will prevail.

7.3.6.8.3 Analytical Artifacts

There is a potential problem when analyzing suspended particles, solutes, and electrolytes. If we could only convert the liquid phase of water to an amorphous glass by cooling the whole sample at a rate of 10^6 K/s, the problem would not exist, as everything in the sample would remain in the same position to within a few nanometers. This level of preservation is only possible in very thin (10- to 20-nm) layers of aqueous suspension.

Using high pressure freezing, it is possible with 200- to 400-µm thick specimens, to obtain amorphous ice on the outer 20 nm of the sample and 50 nm^3 microcrystalline ice crystals in the rest of the specimen. In samples no larger than 1 mm^3, the cooling rate within the specimen produces 80- to 100-nm^3 ice crystals that compromise the structural fidelity for the highest resolution SEM images.

In a 5-mm^3 specimen cooled in melting nitrogen, the ice crystals can be as big as 5 µm^3: the larger the specimen, the slower the cooling and the larger the ice crystals. During sample cooling, the non-aqueous components are swept to the edges of the developing ice crystals where they remain. During subsequent freeze-drying, the so-called ice crystals "ghosts" appear as naked voids in the sample from which the ice has been sublimed. This movement of the non-aqueous components produces artificially high concentrations of material away from their original location that makes high spatial resolution x-ray microanalysis impossible to achieve. This vexed question of the relocation of soluble elements, electrolytes, and non-soluble components within cells and the redistribution in and loss from tissues remains unsolved.

These problems may be partially resolved by using freeze-dried resin-embedded samples, but as Chapter 8 shows, this approach raises its own catalog of problems. The alternative approach of freeze-substitution, discussed in the next section, provides some answers to these questions.

7.4 Freeze Substitution

7.4.1 Introduction

Freeze substitution, strictly speaking, is not a dehydration process per se and should not be used only to remove water from specimens. As the title suggests, the water is first immobilized at low temperatures and then carefully dissolved in chemicals at low temperatures in order to retain both the *structural* and *chemical integrity* of the sample. Freeze substitution plays an important role in the preparation of biological, moist, and wet specimens for both structural and analytical in the SEM (and the TEM and light microscope) in the four following way:

1. Excellent preservation of fine *ultrastructure*
2. Allows in situ analysis of *soluble components* such as electrolytes and soluble inorganic and organic molecules
3. Permits localization of biological *macromolecules* such as enzymes, proteins, carbohydrates and drugs
4. Permits localization of regions of *functional integrity* of biochemical processes such as respiration, neural activity, and photosynthesis

7.4.2 The General Principles of Freeze Substitution

In sharp contrast to the process of freeze-drying discussed in the previous section, where the ice is *removed* from the specimen by a purely physical process and is not *replaced*; in freeze substitution, organic solvents are used to *dissolve* the ice at temperatures low enough to avoid ice re-crystallization.

Freeze substitution is always carried out at temperatures below 273 K. Once the substitution by organic solvents is complete, the sample may either be freeze-dried or critical point dried for SEM. Alternatively, the substituted specimens may be infiltrated in liquid resins, which are then polymerized and either sectioned or fractured for SEM. The resin infiltration and polymerization can either take place at low temperatures or at room temperatures. Freeze substitution is primarily a preparative method for biological samples and bio-pharmaceuticals and food materials in which enzyme activity is an important component. In the last 20 years there has also been a lot of activity to develop techniques to improve the potential for structural preservation.

There are three steps in the preparative procedure that need close attention to ensure that the substitution process results in minimal loss of structure and maximum retention of biochemical activity:

1. Macromolecules should be immobilized in an inert manner to minimize the likelihood of massive supramolecular rearrangement and subsequent denaturation of cellular and enzymatic components. This preserves cell architecture in a more lifelike

form and ensures that the structural integrity of the membranes remains intact and their chemical components are in the correct subcellular compartments.
2. The solvent environment should remain as "water-like" as possible or as polar as possible throughout the entire dehydration and embedding. It is important to maintain the hydration shell surrounding macromolecules and prevent disrupting the chemical ligands and bonds involved in inter- and intra-molecular contact. This factor is probably less optimal for preserving the smaller molecular weight highly water soluble components.
3. Working at low temperatures enhances the stability of biological material during solvent exchange by reducing the amplitude of molecular thermal vibration.

These three important goals are only realized if the polarity, identity, and chemical interactions within and between specimens are retained.

7.4.3 General Outline of the Procedures Used for Freeze Substitution

In terms of practical usage, the various steps in freeze substitution are best integrated into a continuous series of processes that are initiated with sample stabilization and then continue through quench cooling, dehydration, embedding, and finally polymerization. This requires the use of special equipment to ensure that once the sample is quench frozen it remains at low temperatures in an anhydrous environment until the resin embedding is complete. Within this general scheme there are a number of variations depending on the ultimate destination of the samples. There are no "standard" procedures for freeze substitution that provide a general account of the different options that may be followed. Within this general picture the activities of specimen stabilization, cooling, substitution, and sample destination are considered.

7.4.3.1 Specimen Stabilization

The chemical intervention can either take place while the sample is in aqueous phase *prior* to cooling or, as is more usual, incorporated into the substitution fluids that are used immediately *after* the quench cooling.

The procedure and option for sample stabilization are discussed in detail in Chapters 8 and 9, and it is assumed that all forms of chemical stabilization can cause changes to the natural state of the sample. Because of the four different reasons why one should consider using freeze substitution (ultrastructure, soluble components, macromolecular components, and functional integrity), it is not possible to provide hard and fast recipes here.

However, the general approach should attempt to incorporate "mild" stabilization procedures.

7.4.3.1.1 Chemical Stabilization
1. Use low concentrations of glutaraldehyde or acrolein for short periods of time either *before* quench cooling or *after* quench cooling during freeze substitution.
2. Include low concentrations of heavy metal salts such as osmium tetroxide and uranyl acetate into the organic substitution fluids.
3. Use bifunctional cross linking agents such as dimethyl-adipimidates (Tzaphilodou and Mattoupoulos, 1988) and imido-esters and the carbo-diamides (Bullock, 1984) as initial stabilization chemicals.

7.4.3.1.2 Staining
In addition to using heavy metal salts in the substitution fluids, they have also been used to increase the electron scattering in organic materials. Fernandez-Moran (1959) used platinum and gold chlorides, and Heuser (1989) used hafnium chloride as stains. Bridgeman and Reese (1984) examined a whole range of substitution fluids and found that the addition of tannic acid to the substitution fluid enhanced the staining effects of osmium tetroxide and uranyl acetate. Periodic acid, thiocarbohydrazide, and a silver proteinate complex have also be used as staining protocols.

7.4.3.1.3 Cryoprotectants
The situation with regard to using these agents is uncertain, as their use imposes restrictions on the total information that may be obtained from the specimen. There seems to be no *a priori* reason why cryoprotectants should not be used prior to quench cooling for morphological and ultrastructural studies. Their use with immunocytochemical studies of macromolecules and x-ray analytical studies on electrolytes and small molecules remains questionable. As a general rule, cryoprotectants should not used unless their effects are fully understood.

7.4.3.2 Specimen Cooling

These processes are discussed in detail in Chapter 8. The same quench cooling procedures described for specimens that are *frozen-hydrated* and examined and analyzed at low temperatures, apply equally well to samples destined for *freeze substitution*.

7.4.3.3 Specimen Substitution

These processes, which are considered in more detail later involving dissolving the frozen specimen in different types of organic liquids, which may or may not contain additional chemicals to ensure sample stabilization.

7.4.3.4 Specimen Destination

After the water has been successfully substituted in an aqueous specimen, it has a number of different destinations before it may be examined and analyzed in the microscope. Four different options are available because the biggest impediment to examining wet specimens in an electron beam instruments has been removed. The samples are anhydrous:

1. The specimen may be re-frozen, cryo-fractured, or cryo-planed and examined at low temperatures in a microscope fitted with a cold stage.
2. The substituted sample may be critical point dried as described earlier.
3. The substituted sample may be embedded in resins at either low or ambient temperature.
4. The substituted specimen could be freeze-dried.

7.4.4 Low Temperature Refrigerators for Freeze Substitution

Figure 7.14 gives a schematic representation of a freeze substitution refrigerator. The refrigerators must be capable of more than just providing low temperatures. It is possible to build such a refrigerator, but the equipment is now available commercially.

The commercially available freeze substitution refrigerators are sophisticated pieces of equipment. See, for example, the Leica EM AFS2 (Figure 7.15) (www.leica-microsystems.com) and the Baltec-RMC FS-7500 (www.rmcproducts.com) freeze

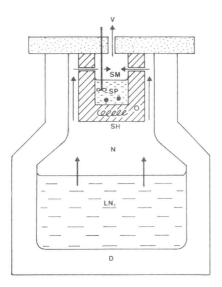

Figure 7.14. Diagram of a freeze substitution container. V = vent, SM = substitution liquid, SP = specimens, SH = metal specimen holder, N = cold nitrogen gas, LN_2 = liquid nitrogen, D = large Dewer (Picture from Echlin, 1992)

Figure 7.15. Leica EM AFS2 portable freeze substitution device situated on a large liquid nitrogen Dewer (Picture courtesy of Leica Microsystems GmbH, Vienna, Austria)

substitution unit. Both pieces of equipment are highly automated and allow both freeze substitution and progressive lowering of temperature (PLT) methods as well as low temperature embedding and resin polymerization.

The following specifications should be considered when either purchasing or constructing a freeze substitution unit:

1. The refrigerator must cover the full temperature range being used, and maintain it for several weeks if necessary.
2. It should be possible to raise and lower the temperature of the specimen in a continuous, controlled, and reproducible manner.
3. Facilities must be available to store, dispense, and exchange all substitution fluids at the appropriate temperature and under strictly anhydrous conditions.
4. The unit should be capable of carrying out the low temperature resin embedding procedures that frequently follow on from the freeze substitution.
5. It should be possible to create an oxygen-free and anhydrous environment during the period of resin polymerization.
6. The equipment should be fitted with the appropriate UV sources and controls for use to polymerize most of the embedding resins.
7. It should be possible to easily manipulate small samples and pieces of equipment that must operate at low temperatures and in an anhydrous environment.

8. The different substitution, stabilization, embedding, and polymerization events should proceed continuously within the same piece of equipment.
9. From the technical point of view, the simplest and most effective way of achieving, maintaining, and varying low temperatures in the 293–173 K range is to use LN_2 as the coolant and electrical power as a source of heat.

The samples are best kept in airtight transparent containers with secure lids that are easy to remove at low temperatures. The containers should fit snugly into holes in a metal block that can be maintained at the appropriate temperatures by cooling with LN_2 and heating with a small electric element. The cooling and heating components should be linked to an efficient programmable temperature controller and thermocouple. The sample cooling block should be placed at the bottom of a well-insulated container, the top of which is fitted with a transparent, tight-fitting lid. There should be sufficient space in the insulated box to hold a slow rotary stirrer and an indirect, low voltage UV source. The interior of the box should be lined with shiny aluminum foil and continuously flushed with a gentle flow of dry nitrogen to create an anhydrous anoxic environment. Finally, there should be sufficient space in the cold box for the various substitution and embedding liquids that should be kept over molecular sieve and in sealed dark glass bottles where appropriate.

7.4.5 Practical Procedures for Freeze Substitution

Ideally, one should only use organic liquids that can dissolve the sample ice below its re-crystallization temperature of ice, i.e., ca. 140 K. There are very few organic chemicals that would dissolve ice at these temperatures, and the process would take far too long to be of practical use. In practice, substantial re-crystallization does not occur in hydrated samples until such temperatures are reached, ca. 190 K, and there are a number of organic liquids that will dissolve ice at these temperatures in a reasonable period of time. Table 7.6 gives properties of the organic liquids that are

Table 7.6. Temperatures at which dehydration and embedding take place for different dehydration and embedding agents

Dehydration agent	Embedding resin	Embedding temperature (K)
Ethylene glycol	K4M	238
Methanol	K4M	238
Ethanol	Epon	293–343
Acetone	Epon	343
Methanol	HM20	223
Ethanol	HM20	223
Acetone	HM20	293

Data from Carlemalm et al., 1982.

most commonly used as substitution media. Robards and Sleytr (1985) give a wider selection together with some eutectic mixtures such as ethylene glycol/water and ethylene glycol/methanol.

Schedules for three of the most popular substitution fluids are given in the following.

7.4.5.1 Acetone

This is one of the most widely used substitution fluids for structural studies and a typical schedule is as follows:

1. Small pieces of quench cooled specimens are quickly placed in an open mesh wire basket which is immersed in liquid nitrogen.
2. The basket is quickly transferred to anhydrous acetone at 193 K, with or without stabilization chemicals, and kept at this temperature for 2–3 days.
3. The cold substitution fluid should be replaced several times with fresh cold anhydrous acetone from an air tight bottle containing pellets of Linde molecular sieve. The molecular sieve is an effective drying agent, but care should be taken not to shake the bottle as small pieces of the drying agent may contaminate the sample.
4. Following substitution at 193 K the sample is allowed to warm up to 233 K over a 12- to 24-h period.
5. At this stage, the sample can either be transferred at 233 K to a low temperature embedding schedule or slowly allowed to warm to room temperature and, following several rinses in anhydrous acetone, either dried or embedded in an appropriate resin at room temperature.

7.4.5.2 Diethyl Ether

This method is particularly useful where the principle objective is to retain the original location of water soluble substances and electrolytes. A typical schedule is as follows:

1. Quench frozen samples kept in liquid nitrogen are quickly placed in anhydrous diethyl ether at 193 K for 20 days, with or without stabilization chemicals, followed by 1 day at 233 K. Alternatively, the substitution may be carried out at lower temperatures over a period of several weeks.
2. The sample may either be transferred at 233 K to a low temperature embedding schedule or allowed to slowly warm to room temperature and, following several rinses in anhydrous diethyl ether, either dried or embedded by room temperature methods.
3. Water has a very low solubility in diethyl ether and it is important to use this substitution fluid in conjunction with a drying agent.

7.4.5.3 Methanol

This has the advantage that it can be used to substitute samples with relatively high amounts of water. It is faster than acetone and can be used for both immunocytochemical and structural studies. A typical schedule follows:

1. Quench frozen samples kept in liquid nitrogen are quickly placed in anhydrous methanol at 178 K, with or without stabilization chemicals, for 24–28 h, after which the mixture is slowly warmed to 273 K over a 24-h period. Alternatively, the substitution may be carried out at lower temperatures over a period of several weeks.
2. The sample may either be transferred at 273 K to a low temperature embedding schedule or allowed to slowly warm to room temperature, and following several rinses in anhydrous methanol, either dried or embedded by room temperature methods.

Practical suggestions:

1. Do not hurry the period of substitution.
2. Temperatures should be slowly increased using a continuous temperature rise rather than stepwise increases.
3. Only use anhydrous substitution fluids and keep them dry with drying agents.

7.4.6 Specimen Handling Procedures

One of the major difficulties with freeze substitution (and low temperature embedding) is to handle small specimens at low temperatures and under anhydrous conditions. The specimens are best placed in small wire baskets that can easily fit into the containers that hold the freeze substitution fluids. Disposable plastic transfer pipettes can be used to change the substitution fluids.

7.4.7 Advantages and Disadvantages of Freeze Substitution

7.4.7.1 Advantages

It is generally accepted that this is one of the best ways of preserving the functional ultrastructure of biological, moist, and wet materials, although there are no good criteria against which this may be judged. The most commonly used objective and quantifiable criteria for judging the quality of material preparation are as follows:

1. Cell membranes should appear smooth, continuous, and in sharp contrast.
2. The size and distribution of cellular components and compartments should remain unchanged.

3. The cytoplasm should appear dense and slightly granular, which would indicate that little extraction had taken place.
4. There should be no large "ice crystal ghosts," which would indicate poor initial quench cooling and possibly, subsequent re-crystallization.
5. There should be no transparent halos around organelles that would indicate shrinkage.
6. There should be no separation between the specimen and any embedding resin.
7. In any subsequent immunocytological studies, the specimen should retain its antigenicity.
8. In any subsequent x-ray microanalytical studies the spatial distribution and concentration of electrolytes and soluble elements should remain unchanged.

These eight rather demanding criteria apply equally to SEM and TEM samples.

7.4.7.2 Disadvantages

The use of organic solvents during both the freeze substitution and any subsequent embedding may present a problem with dissolved fats and lipids and their loss from the specimen. This loss is significantly decreased at low temperatures and when careful sample stabilization has been carried out.

The scientific literature contains a large number of papers that show, invariably for one particular species, that freeze substitution is a reliable preparative technique for biological moist and wet specimens. However, what works for one particular sample may not necessarily work for another, and each new sample will probably need subtle changes to the preparative protocol.

There are two other low temperature procedures for removing water from samples, one of which does not involve freezing the samples, and the other that does not involve freezing samples at very low temperatures. These two procedures are generally used in association with light microscopy rather than electron microscopy, but they are included here as they may have applications to particularly sensitive biological samples.

7.5 Low Temperature Dehydration by Progressive Lowering of the Temperature

The cool but *unfrozen* sample is slowly cooled further during passage through increasing concentrations of dehydrating agents. Ethylene glycol was initially used as it is a very polar solvent and one of the weakest denaturing organic solvents at below 273 K. A careful study by Carlemalm et al. (1982) on a number of different proteins showed that although ethylene glycol caused minimal *denaturation* of molecular components, it may cause *disruption* of

Table 7.7. An experimental schedule for the progressive lowering of temperature dehydration using ethanol or acetone

Ethanol	Acetone
30% for 30 min at 273 K	30% for 30 min at 273 K
50% for 60 min at 258 K	50% for 60 min at 258 K
70% for 60 min at 238 K	70% for 60 min at 238 K
95% for 60 min at 238 K	100% for 60 min at 243 K
100% for 60 min at 238 K	100% for 60 min at 243 K
100% for 60 min at 238 K	

Table 7.8. Melting point and water containing capacity at 193 K of solvents used in freeze substitution

Compound	Melting point (K)	Percentage of water dissolved at 193 K
Acetone	177.6	2.5
Diethyl ether	156.5	0.8
Di-methyl ether	134.5	
Ethanol	155.7	16
Methanol	179.1	32
Propane	83.3	0.1

the macromolecular structure. Any polar dehydrating solvents must be compatible and miscible with the resins that are to be used to embed the specimens. Tables 7.7 and 7.8 show some of the properties of different dehydration and accompanying embedding agents and gives a procedure to progressively dehydrate a sample using ethanol and acetone.

7.6 Isothermal Freeze Stabilization

Although, again, not strictly a process associated with dehydration, it is a procedure associated with studies on the distribution of intracellular and extracellular ice and the effects of freezing damage in hydrated specimens at and around 273 K.

The amount, formation, and distribution of ice in freezing biological systems is an important factor in understanding the processes of low-temperature injury and preservation in such activities as agriculture, food technology, pharmaceutical agents, and tissue preservation. These types of samples are particularly sensitive to ice crystal damage during any manufacturing procedure that involve subzero temperatures. With these types of specimen it is not so much the structural damage that is of concern but the biochemical changes that arise as a consequence of freezing living material.

The process of isothermal freeze fixation involves chemical stabilization of samples in the presence of cryo-protectants at temperatures just below 273 K under conditions of constant temperature and water activity. The composition of the stabilization solution is carefully adjusted so that its freezing point corresponds to the temperature to which the sample has been cooled in the first place. This ensures that the stabilization chemical(s) diffuses through the sample under equilibrium conditions without disturbing either the configuration of any naturally formed ice crystals or the state of sample hydration. Once the stabilization is completed, the sample may be thawed and processed for study in the microscope. The technique has been used at temperatures between 252 and 273 K using different solutes in the stabilization liquids to ensure that it is isotonic with the concentrated solute phase in the tissue.

Isothermal freeze stabilization has the advantage that it preserves the structure of ice at the temperature at which it was formed. In contrast, techniques such as freeze-drying, freeze substitution, and freeze fracture replication, which are based on non-equilibrium cooling, all require the sample to be cooled to low temperatures. This can seriously affect the configuration and morphology of ice formed under near equilibrium conditions at higher temperatures, which in turn alters the structural appea-rance of the tissues.

Hunt (1984) provided an excellent review of the procedures of isothermal freeze substitution and showed how it is possible to design freeze stabilization procedures for temperatures 253–273 K. Thermal equilibrium is ensured by maintaining both the sample and the stabilization chemicals at a constant temperature. The addition of a small amount of finely divided ice is added to the freeze stabilization mixture. These act as foci for any ice crystal growth and ensure that minor fluctuations in temperature do not affect the ice phase in the sample.

The distinct advantage of isothermal stabilization over freeze substitution carried out at the same temperatures is that the ice matrix remains essentially undisturbed during stabilization. Most of the studies using this method have used the light microscope, but there is no reason why the same sort of material should not be examined in environmental and variable pressure SEM. These instruments have the specimen chamber and the column divided into different pressure regions separated by pressure limiting apertures. This allows hydrated samples and sample containing ice in their natural state at pressures between 10–2,500 Pa (0.1–20 Torr) and temperatures as low as 263 K.

7.7 Suggested Dehydration Procedures for the Six Different Types of SEM Specimens

There are two problems with the removal of water from specimens. The first is the high surface tension at the water/air interface, which can rise to as high as 20–100 MPa (2,000 psi) during the final phases of air drying and damage the sample. Small soft specimens are damaged more than larger hard specimens. The second problem is related to the fact that water is a near universal solvent and in some investigations it is important to know the natural location of the dissolved and suspended materials. As the water is removed the dissolved components move from their original location in the sample.

7.7.1 Metals, Alloys, and Metallic Materials

These types of sample are sufficiently robust that they may be slowly air dried by placing them in a dust free environment at room temperature or for a short period of time in a poor vacuum 400 Pa (30 Torr) of dry argon gas. Alternatively, they may be dehydrated with jets of dry nitrogen or air, jets from cans of compressed air such as Spraydustoff™, or simply by using a warm (333 K) domestic hair dryer. Avoid using a heat gun, which can damage the sample. These procedures dry wet specimens and specimens that have had a final rinse in methanol, ethanol, or acetone. Care must be taken to use clean air to avoid leaving a layer of dust and particles on the surface and volatile organic materials should be evaporated in a fume cupboard.

7.7.2 Hard, Dry, Inorganic Materials

These samples may be dried in the same way used for metal samples. Care has to be taken with small samples, which can be blown away by pressure jet of clean air and gases.

7.7.3 Hard and Firm, Dry Natural Organic Material

The same drying procedures used for the two previous types of specimens can be used here, provided the specimens are sufficiently robust. Samples such as papers, fabrics, furniture, and naturally and kiln dried wood may be dried using the methods described in the two previous samples. Passive chemical dehydration may be used on soft samples provided they are not affected by the drying agents. Critical point drying can be used on delicate specimens that have undergone some form of chemical stabilization.

7.7.4 Synthetic Organic Polymer Material

Most polymers and polymeric materials are naturally dry; if they are wet and sufficiently robust, they may be air dried. Other polymers such as membranes and emulsions contain varying amounts of water and the surface tension effects of air drying may destroy or distort the specimens. Sawyer et al. (2008) suggest that wet porous membranes be cut into thin strips and placed into a jar containing a small amount of water containing an appropriate embedding resin. The sample is placed in a low vacuum c.a. 2,500 Pa (20 Torr) for several hours to remove the water, which is then replaced with fresh resin. Other methods of dehydration might include critical point drying and freeze-drying. Chemical drying may be used provide none of the dehydrating chemical damage or dissolve the organic polymers.

7.7.5 Biological Organisms and Materials

Air drying is generally not an option. Many of the other methods described in this chapter may be used to dehydrate this broad group of samples.

7.7.6 Wet and Liquid Materials

The very nature of this type of sample precludes any discussion about dehydration. There are only two ways these samples may be studied in the SEM: in the variable pressure or environmental instruments where the water and liquids are studied at a poor vacuum, or in a low temperature SEM where the water and liquids are converted to a solid phase that is examined at high vacuum.

8

Sample Stabilization for Imaging in the SEM

8.1 Introduction

This chapter considers the ways by which it is possible to stabilize samples so they may withstand the alien environment inside the SEM in order to produce good images and, as shown in the next chapter, reliable analytical data.

The process of stabilization is frequently referred to as *fixation*. This is an ill-defined term; it is more appropriate to use the term *stabilization* as this more closely describes what is hoped to be achieved with the sample. Some samples need little or no stabilization as they are electrically conductive, radiation resistant, and unaffected by the high vacuum and arid environment inside the microscope column. In contrast, a wide range of samples are very susceptible to all four characteristic features of the inside of the SEM and, consequently, need careful stabilization. The total process of sample stabilization must also include charge elimination. This important topic is discussed in Chapter 11.

One of the many advantages of the SEM is that the interaction of the primary beam with the sample can give rise to a number of different signals. Some of these signals are more appropriate to providing chemical information about the sample and are considered in the next chapter. Topographic (image) contrast mechanisms provide the high quality 3D details such as the shape, size, and surface texture of specimens, and are the basic imaging processes of the SEM. The images are generated either by secondary electrons (SE), which have a very shallow (a few nanometers) escape depth from the specimen surface or by the backscattered electrons (BSE), which are generated much deeper (many nanometers) in the specimen.

These two signal mechanisms are not discussed in detail here; those readers who are not familiar with the SEM and its modes of operation should read the appropriate chapters in the recent textbook by Goldstein et al. (2004). The secondary electrons (SE) have energy of less than 50 eV and arise primarily from the sample surface. In contrast, the backscattered electrons (BSE) have energy as high as the incoming electron beam, i.e., several kV, arise from several micrometers below the sample surface.

The processes of stabilization to be considered here aim to strengthen the molecular structure of the sample, retain its chemical identity, either eliminate or inactivate any liquids, and if possible make the sample more resistant to radiation damage. It is convenient to consider these processes of stabilization under two general headings:

1. Sample stabilization using chemical methods at ambient temperature to ensure high quality representative images of the specimen. Although these procedures are invasive, they can provide a wealth of excellent images.
2. Sample stabilization using non-chemical methods at ambient and low temperatures to ensure high quality images and chemical information about the specimen. In contrast to the previous processes, the stabilization procedures are essentially non-invasive.

The central feature of these two approaches is to ensure that the sample structure and chemical identity remain as close as possible to their original state and identity. The aim should be to use the minimally invasive procedure necessary to stabilize the sample. The processes used draw heavily on the many methods developed for both light and transmission electron microscopy.

8.2 Sample Stabilization for Imaging in the SEM Using Chemical Procedures at Ambient Temperature

The stabilization procedures are considered under the six categories of samples used elsewhere in this book.

8.2.1 Metals, Alloys, and Metallized Specimens

Provided the specimens are chemically stable, electrically conductive, un-affected by the electron beam, and have been properly polished, etched, re-polished, and thoroughly cleaned, there is no need for any sample stabilization. Some metals oxidize very quickly and the only way to look at a non-oxidized surface is to plasma etch the surface in a high vacuum device attached to the side of the SEM. This type of problem arose during early attempts in my laboratory to sputter coat samples with chromium. Using a plasma etching device in a high vacuum chamber linked via an air lock to the microscope column, it was possible to first remove the metal surface oxide layer with high purity argon plasma and then sputter coat the sample with pure chromium. The cleaned and coated surface was kept at high vacuum and transferred directly to the microscope column.

In some cases the metal is part of a sample that is made up of one or more of the other five categories of specimens discussed earlier. For example, many food packaging materials consist of a thin plastic layer coated with a very thin film of aluminum. The first approach would be to try and physically separate the two layers and study then separately. Long practical experience by the author in British pubs has shown that this is possible with most samples of potato crisps (chips) packages. If this is not possible, then a small piece of the metallized foil should be removed and attached to a metal stub using conductive paint and the sample examined at low voltage SEM.

8.2.2 Hard, Dry, Inorganic Materials

These samples, like those from the previous group, need little or no stabilization. Their main problem with SEM and x-ray microanalysis, is that they are all poor electrical conductors. This problem is addressed in Chapter 11. Provided the samples are dry, they are usually chemically stable. Some samples such as clays, soils, mud, sands, and suspended particles contain varying amounts of organic and inorganic liquids, primarily water. The liquids must either be removed before carrying out ambient temperature SEM or converted, at sub-zero temperatures, to a solid state before performing low temperature SEM. These procedures are discussed later in this chapter.

If the organic components are of prime importance, the sample should first be thoroughly washed in de-ionized water and then in an appropriate solvent such as acetone, to remove any organic components. The washed sample must be dried before examination in the SEM and the organic and inorganic washings should be dried and studied separately. These procedures of sample stabilization are very invasive because the two washings, although they may successfully separate the soluble organic and inorganic materials, leave no trace of the location these materials in the sample. If the spatial location of the soluble components is important, the intact sample must be quench cooled and examined and analyzed at low temperatures.

8.2.3 Hard, Dry, Natural Organic Materials

These natural dry materials are non-conducting and, when wet, less chemically stable than the previous two types of samples. It is virtually impossible to improve their stability by non-invasive methods. Although the near anhydrous state of these specimens precludes the use of enzyme based reactions, it is possible to use some histochemical procedures to enhance both their conductivity and increase the coefficient of backscattered electrons. Many of these natural organic materials are porous and their backscattered electron coefficient may be increased by first marinating the specimen in solutions of heavy metals salts, such as silver or osmium, after which the sample is briefly washed with de-ionized water and air dried at 293–303 K. Because these techniques have been primarily designed to increase sample conductivity, details of the protocols are discussed in Chapter 11.

8.2.4 Synthetic Organic Polymer Materials

This group of materials is characterized by being somewhat chemically unstable, non-conductive, and beam sensitive. This is the first of the six groups of specimens that require special attention to sample stabilization, and this topic has been admirably considered in the recently published comprehensive third edition of the book by Sawyer et al. (2008).

A synopsis of some of their stabilization procedures is given here. Sawyer et al. (2008) are full of practical experimental procedures and well-tried recipes for polymer stabilization and staining.

The stabilization procedures center on minimizing the radiation effects and improving sample surface conductivity and signal-to-noise ratio. Ionization radiation exposure and thermal effects damage polymers by breakage chemical bonds, mass loss, reduction of crystallinity, and the formation of volatile material in the high vacuum of the SEM. These deleterious effects may be minimized, but not entirely removed, by using the minimum beam energy necessary to provide the maximum information about the sample. Low voltage and low beam current, low temperature, and variable pressure microscopy, all help to minimize bean damage. The procedures to improve sample conductivity are discussed in Chapter 11, but some of the methods used to stabilize the specimens and improve their signal-to-noise ratio are discussed here.

Most polymers, in common with biological materials, are composed of low atomic number elements and hence exhibit a low signal to noise ratio. The specimen contrast may be enhanced by staining with high atomic number compounds that are either chemically or physically incorporated into the polymer. Unlike biological stains that usually have highly specific staining sites, there is little differential contrast or site specificity in most polymers. Most of the stains used with polymers have a positive charge and the region of interest is stained dark either by chemical interaction or selective physical absorption. In contrast, negative stains, which can be used to stain small size polymers, such as latex or emulsions mounted on a flat surface, tend to enhance the surrounding support matrix and leave the polymer unstained.

Staining of polymers has been reviewed in detail in the book by Sawyer et al. (2008) and only a general outline is considered here. There are two methods that can be used to stabilize some soft polymers and a number of other methods that can be used to stain particular chemical groups.

8.2.4.1 Chlorosulfonic Acid

This light element compound stabilizes and cross-links the amorphous materials in semi-crystalline polymers such as polyolefins by incorporating chlorine and sulfur on the lamella surfaces. The stabilized material may be post-stained with uranyl acetate. The procedure are given in Table 8.1 and an image of a stained specimen is shown in Figure 8.1.

8.2.4.2 Ebonite

The multiphase polymer specimen shown in Figure 8.2 is material from an automobile tire that contains rubbers, some plastic

Table 8.1. A staining procedure to show the lamella structure of polyethylene

1. The sample is treated with chlorosulphonic acid for 6-9 h at 323K. (Caution: this chemical is highly corrosive and potentially explosive.)
2. Rinse the stained sample, first in concentrated sulphuric acid and then wash in water.
3. Dry the sample and embed it in an epoxy resin.
4. Cut thin sections which are then post-stained in a 0.7% aqueous uranyl acetate for 3 h.

From Sawyer et al., 2008.

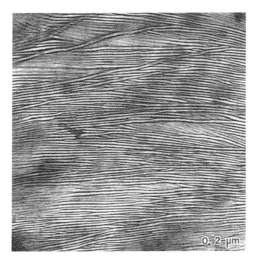

Figure 8.1. Linear polyethylene stained with chlorosulfonic acid (Picture courtesy of Sawyer et al., 2008)

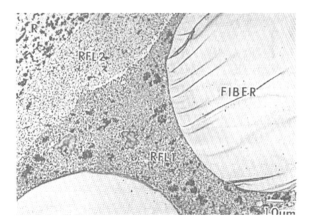

Figure 8.2. Sectioned industrial cord prepared by the ebonite method. R = rubber and REL = adhesive (Picture courtesy of Sawyer et al., 2008)

Table 8.2. The ebonite reaction hardening procedure for rubbers associated with tire cord

1. A mixture is made up as follows in a fume cupboard:
   90 parts molten sulphur.
   5 parts N,H-dicyclohexyl-2benzo-thiazolylsulphenamide.
   5 parts zinc stearate.
2. Small pieces of a sample, such as tire cord, are placed in the mixture for 8 h at 393K.
3. The samples are removed, excess surface material scraped off, and the sample placed in a 393K oven to removed any excess ebonite.
4. The treated sample is embedded in an epoxy resin and sectioned.

From Sawyer et al., 2008.

coatings, and oriented fibers. Such samples are difficult to stain with a single agent. The ebonite method may be used to uniformly stabilize and harden polymers that may be subsequently stained for examination in both the TEM and the SEM by a process referred to as vulcanization, which is shown in Table 8.2. *Caution:* The chemicals are poisonous and the processes must be carried out in a fume cupboard using gloves.

The previous two methods of stabilization are usually followed by one or more heavy metal chemical procedures that are designed to stain particular chemical groups in the polymer and, at the same time, increase the signal to noise ratio and the electrical conductivity of the sample. Some suitable staining methods are shown in the following.

8.2.4.3 Osmium Tetroxide (OsO_4) Z = 76

This very toxic compound can be applied to samples either as a vapor or an aqueous solution. Osmium tetroxide reacts with a wide range of carbon–carbon double bonds and thus can be used as a general stain for a wide range of materials. Figure 8.3 is of an osmium stained polypropylene sample.

8.2 Sample Stabilization for Imaging in the SEM Using Chemical Procedures

Figure 8.3. Stretched polypropylene stained with osmium (Picture courtesy of Sawyer et al., 2008)

Table 8.3. Osmium tetroxide vapour staining of unsaturated polymers

1. A small glass ampule of OsO_4 crystals is carefully scored at the edge and opened and immediately placed in a well sealed glass container.
2. Thin sections of the polymer, stabilized if necessary, are placed on a rack above the OsO_4 in the sealed container.
3. The staining can either be carried out at room temperature or accelerated by placing the sealed container in a beaker of hot water at 323K.
4. The staining time is very variable and depends on the temperature and degree of saturation of the polymer. For example, a section of an unsaturated rubber or a semi-crystalline polymer would be sufficiently stained after 1–2 h at 323K.

From Sawyer et al., 2008.

Some polymers may need pre-treatment with alkaline saponification before they will take up the stain. Osmium tetroxide not only acts as a stain for polymers such as unsaturated rubbers, latex particles, and elastomers, but it also tends to harden the materials and makes it easier to cut thin sections for microscopy. Two osmium staining procedures for polymers are given in Tables 8.3 and 8.4. *Caution:* The vapor and dissolved phases of osmium tetroxide are poisonous and the processes must be carried out in a fume cupboard using gloves.

8.2.4.4 Ruthenium Tetroxide (RuO_4) Z = 44

This is a powerful oxidizing agent and is a more effective surface stain than osmium tetroxide. It can be used either as a 1%

Table 8.4. Schedules of different osmium tetroxide solutions used to stain polymers

1. 10ml of a 4% osmium tetroxide solution is made by dissolving crystals in pure water.
2. Thin sections of polymers with carbon-carbon double bonds are immersed for seven days in the 4% OsO_4 solution after which they are washed in water and dried.
3. The staining of polymer hydroxyl groups may be enhanced by first treating with alkaline saponification.
4. Thin films of melt crystallized polyamines can be stained by a mixture of equal parts of 30% aqueous formaldehyde and 1% OsO_4.
5. Rubber modified epoxy resins may be stained using OsO_4 dissolved in tetrahydrofurane.
6. The staining times and schedules for polymers are very variable.

From Sawyer et al., 2008.

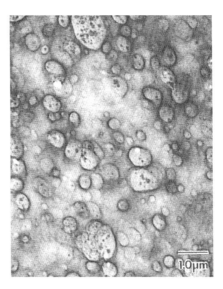

Figure 8.4. A nylon blend with a saturated elastomer stained with ruthenium tetroxide (Picture courtesy of Sawyer et al., 2008)

aqueous solution or a vapor. The chemical acts by oxidizing aromatic rings, ethers, alcohols, and amino groups and is generally believed to be a good staining agent for rubbers, latex, resins, nylon, polystyrene, and polyethylene. For example, thin polymer films can be vapor stained when suspended over an aqueous 0.5% RuO_4 solution for 30 min. Figure 8.4 is a blend of nylon stained with ruthenium tetroxide.

The procedures discussed in the preceding for osmium tetroxide and ruthenium tetroxide are generally use the two elements as an aqueous solution. A recent paper by Ellis and Pendleton (2007) describes a simple and effective method for using the vapor of these two elements to stabilize packing foam. Their vapor stabilization technique has also been used on biological specimens.

8.2.4.5 Phosphotungstic Acid (PTA) ($H_3[P(W_3O_{10})_4] \cdot 14H_2O$) Z = 74

Surface functional groups, such as hydroxyl, carboxyl and amines, of some polymers, interact with this chemical (PTA). For example, nylon-6 fibers can be stained by first soaking then in a 10% aqueous salt solution followed by a 1–5% aqueous solution of PTA. Alternatively, a 2% solution of PTA and 2% benzyl alcohol can be used as a stain for a number of different polymers. Figure 8.5 is of latex particles stained with PTA.

8.2.4.6 Silver Sulfide (AgS) and Silver Nitrate $AgNO_3$ Z = 47

Silver salts, which have long been known as effective stains for biological polymers such as hair, wool, and regenerated cellulose, also can be used to stain some synthetic polymers. Aqueous solutions of the two silver salts have an affinity for poly-vinyl-alcohol, acrylics, and polyesters groups. Aramide fibers made from linear polymers containing recurring amino groups may be stained with silver salts. Care must be taken when staining polymers using silver sulfide as the specimen must first be treated with gaseous hydrogen sulfide at 1.38 MPa (200 psi) for 16 h at 293 K, washed in alcohol, and immersed in a 5% aqueous solution of silver nitrate for 3–4 h at 293 K.

8.2.4.7 Specialized Preparative Methods

In addition to the different single chemical methods mentioned in the preceding, it is possible to use mixtures of these materials; Figure 8.6 shows an example of a mixed stain application.

Compounds of mercury and iodine can be used to stain polymers. Some polymers, such as latex suspensions, emulsions, and

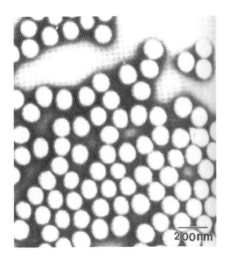

Figure 8.5. Poly(ethyl) acrylate latex particles negatively stained with 2% aqueous phosphotungstic acid (PTA) (Picture courtesy of Sawyer et al., 2008)

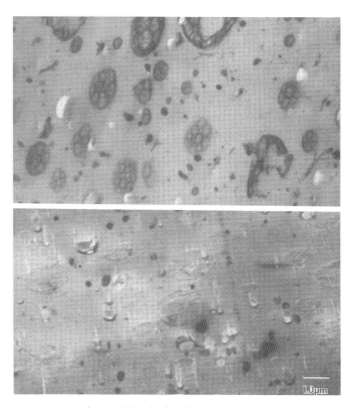

Figure 8.6. A copolymer blend of high impact polystyrene–polyethylene and ethylene. The *upper image* is of material stained with OsO_4 to reveal the butadiene rich structure of the high impact polyethylene. The *lower image* is of material stained with RuO_4 to show the regions rich in polyethylene (Picture courtesy of Sawyer et al., 2008)

adhesives, either contain water or are dispersed in an aqueous phase. Although the procedures for dealing with water are discussed later, it is important to know whether the water is an integral part of the polymer or only acts as a supporting medium. Sawyer et al. (2008) provide a comprehensive list of specific polymer functional groups and how they may be stained for electron microscopy (Table 8.5).

8.2.5 Biological Organisms and Materials

This group of specimens and the following group of wet samples invariably need extensive and careful stabilization before they may be studied properly in the SEM. We need only to consider our own bodies as an example to appreciate the complexity and chemical frailty of biological material. We are composed primarily of light elements, most of which are organized into a mixture of periodic and aperiodic structures. These structures range from simple molecules and macromolecules to complex heteropolymers, all bathed in an aqueous solution of ions and electrolytes.

Table 8.5. Different functional groups found in polymers with examples of how they may be stained for electron microscopy

Function group	Example	Stains
-CH-CH-	Saturated hydrocarbon i.e. PE, PP and HDPE	Chlorosulphonic Acid, Phosphotungstic Acid, Ruthenium Tetroxide
-C=C-	Unsaturated hydrocarbons i.e. Polybutadiene rubbers	Osmium tetroxide, Ebonite, Ruthenium tetroxide
-OH & -COH	Alcohols and aldehydes	Osmium tetroxide, Ruthenium tetroxide, Silver sulphide
-O-	Ethers	Osmium tetroxide, Ruthenium tetroxide
$-NH_2$	Amines	Osmium tetroxide, Ruthenium tetroxide
-COOH	Acids	Hydrazine then Osmium tetroxide
-COOR	Esters	Hydrazine then Osmium tetroxide, Phosphotungstic Acid, Silver sulphide, Methanolic NaOH
$-CONH_2$ & -CONH	Amides	Phosphotungstic Acid, Tin chloride
Aromatics		Ruthenium tetroxide, Silver sulphide, Mercury trifluoroacetate

From Sawyer et al., 2008.

We are thermodynamically unstable, live at ambient temperatures and pressures, and very sensitive to ionizing radiation.

A simple summary will help to remind us about the problems we face when examining biological material in electron beam instruments. The pressure inside a modern electron beam instrument is less than 1 ten millionth of an atmosphere. It is extremely dry inside the microscope and the interaction between the electron beam and the sample may result in a considerable rise in local temperature. There is a high flux of ionizing radiation and, finally, the highest signal-to-ratio necessary for image formation depends critically on scattering from high-atomic number elements. The inside of an electron microscope is undoubtedly an alien environment for living material.

There are really only two approaches to biological sample preparation. We can either modify the environment inside the microscope column or stabilize the specimens and make then sufficiently robust to withstand the environment inside the microscope. It is not proposed to discuss the first option as the various procedures are considered in detail in Chapters 5 and 10 of the recent book by Goldstein et al. (2004). The second approach of stabilizing and modifying the specimen so that it retains its integrity and remains stable, is considered here.

8.2.5.1 The Development of the Specimen Stabilization Procedures

Scanning electron microscope sample stabilization has developed from the procedures used in transmission electron microscopy, which in turn were modified from the methods used for light microscopy. This is a vast topic; it is impossible here to provide methods for *all* specimens, although a number of generic procedures are suggested.

There are two general approaches to biological sample stabilization: the invasive chemical procedures that are discussed here and the much less invasive low temperature stabilization procedures that are considered in a later section of this chapter. The end product of biological material stabilized for microscopy and analysis is frequently very different from the original biological specimen. The interpretation of the results is akin to the way paleontologists work out the life processes of dinosaurs from 165 million year old rock samples or for us trying to calculate the muscle activity of a steer from a cooked steak.

8.2.5.2 Stabilization by Immersing the Sample in Chemically Invasive Liquids

Immersion is the simplest way to stabilize specimens. The sample should be no larger than 1 mm^3 or, better still, thin slices 0.2–0.5 mm thick. The stabilization is best carried out in glass vials 20 mm in diameter and 30 mm deep with tight fitting lids. The amount of stabilizing fluid should be 20 times greater than the estimated volume of the samples. The specimen should be gently rotated during the stabilization process, which generally should be carried out in a fume cupboard.

For additional information on recipes for biological sample stabilization procedures reference may be made to the recent publications by Bozzola and Russell (1999), Dashek (2000), Glauert and Lewis (1998), Gupta (2000), Hayat (2000), and Maunsbach and Afzelius (1999). Current information on all aspects of specimen stabilization may be found in the different scientific journals discussed in Chapter 13.

The chemical procedures commonly used are based on precipitation, denaturation, and cross-linking, which immobilize the highly mobile solubilized state of cytoplasm, cells, and tissues. This approach has been used for electron microscopy for more than 75 years and has provided a wealth of structural detail. It is, however, now accepted that the fidelity of the fine details have been achieved at the expense of most of the ions and electrolytes and many of the low-molecular-weight molecules such as sugars and amino acids are irretrievably lost from the sample. Thus, our knowledge of biological ultrastructure is based on the structural configuration of the large macromolecules such as proteins, carbohydrates, lipids, and amino acids, which remain after these invasive chemical methods.

However, this bleak view of biological ultrastructure can be ameliorated provided that a number of important principles are borne in mind when formulating and carrying out stabilization procedures.

8.2.5.3 General Guidelines to Formulating Stabilization Recipes

1. In order to examine, either the natural surface or a surface that has been carefully exposed, it is important to ensure they are properly cleaned prior to (and during) stabilization (see Chapter 10).
2. Ensure that the stabilization solutions are as close as possible to the tonicity of the cell or tissue fluids.
3. The term stabilization (fixation) solution consists of all the components of the aqueous buffer, balancing ions and electrolytes, and non-electrolytes such as sucrose in addition to the stabilization chemical.
4. The osmolarity of the stabilization solution is equally important in obtaining satisfactory preservation as the actual concentration of the stabilization chemical. As a general rule, this solution should be slightly hypertonic if the sample is to be sectioned after stabilization.
5. The time needed for satisfactory stabilization is usually shorter with specimens in hypertonic solutions and for those with small dimensions than with specimens that need to be hardened. A balance has to be struck between the longer times required for sample stabilization and the much shorter times that will minimize sample extraction.
6. The pH of the stabilization solution should be as close as possible to the natural environment of the cell and tissue fluids. Plant materials are usually slightly acidic (pH 6.8), whereas animal tissue is slightly alkaline (pH 7.4).
7. Care should be taken with the choice of buffer, as it is important that their effective capacity is within the range of the pH of the specimen.
8. Stabilization of most animal samples is carried out at 277 K; slightly higher temperatures (288 K) may be used for plant material. The best way to judge the stabilization temperature is to mimic the temperature at which the organism lives.
9. The advantage of low temperature stabilization is that the natural biochemical processes are slowed down and ensures that the whole procedure is slowed down. This advantage has

to be balanced against the disadvantage that long stabilization will encourage relocation and extraction of the soluble tissue's components.

8.2.5.4 General Plan of Action

There is no single perfect chemical stabilization recipe, but instead of adopting a hit and miss recipe for a procedure, it is important to bear in mind the following rules when preparing a stabilization recipe:

1. Have a clear understanding of the structural information it is hoped to obtain by using the SEM and/or the x-ray microanalyzer.
2. First assemble as much structural and chemical information as possible about the biological material.
3. Carry out a literature search to find details of any previous SEM studies of the same specimen.
4. Obtain information for the following values:
   a. *The natural pH of the sample.* This is easy to obtain by taking a small amount of liquid from the sample and measuring it in an appropriate meter.
   b. *The osmotic value of the sample.* This is the pressure that must be exerted in a solution containing a given concentration of solute separated from a sample of the pure solvent by a membrane permeable only to the solvent. The whole stabilization solution should be isotonic to the osmotic pressure of the sample to ensure that it neither shrinks nor swells.
   c. *The ambient temperature of the living specimen.* It is presumed that this is the temperature at which the specimen is optionally functional. At temperatures below ambient, the physiochemical living processes may be compromised and the speed of chemical processes slowed down. At temperatures higher than ambient, these same processes may be speeded up; however, raising or lowering the specimen temperature during stabilization has some advantages. Higher temperatures enhance the diffusion rate of the stabilization liquid and the speed of its chemical reactions. Lower temperatures decrease chemical extraction and autolysis from the specimen.
   d. *The permeability of the specimen.* It is generally accepted that because chemical stabilization is an invasive procedure, it should be achieved as quickly as possible. For example, plant cell walls are more permeable than an insect's cuticle; small pieces of carefully dissected tissue are more permeable that the intact organism.
   e. *Toxicity.* Chemical stabilization is an invasive and toxic process that kills living material. Wherever possible, cell and tissue death should be slow and gentle; for this reason

any stabilization protocol should avoid using very toxic materials early in the procedure to avoid gross distortion of the sample.

Bearing these cautionary warnings in mind, Tables 8.7–8.12 for different types of biological specimen are pragmatic introductory recipes to initiate the stabilization of biological specimens prior to image acquisition in the SEM.

8.2.5.5 A General Protocol for Biological Material

The procedure in the Table 8.6 is an amalgam of the general suggestions in the literature for stabilization protocols. Experience has shown that this type of stabilization schedule will provide satisfactory images of most types of biological specimens, but should be considered only as a starting point that may well need subsequent fine tuning to obtain a perfect result.

8.2.5.6 A General Protocol for Plant Material

Plant material ranges from dry, hard, wood, soft photosynthetic tissues to single-cell algae and fungi. They are characterized by a variable thickness and permeable cellulose cell wall that maintains the shape of the cell. The single layer cell membrane and

Table 8.6. A general schedule for stabilizing biological samples by immersion

1. Cut the sample into 1-2mm^3 cubes or 250µm thick sections.
2. Prepare a primary stabilization solution of between 0.5 to 4% glutaraldehyde (OCH-CH$_2$-CH$_2$-CH$_2$-CHO), 0.5 to 2% formaldehyde (HO-CH$_2$-OH) with a 2–3mM calcium chloride (CaCl$_2$) in either:
    a. A 0.1M cacodylate (Na(CH$_3$)$_2$AsO$_2$.3H$_2$O) buffer (100ml of a 0.2M aqueous sodium cacodylate solution plus 1.6ml of N HCl and make up to 200ml with distilled water) pH 7.2.

Or:

    b. A 0.1M PIPES (piperazine-N,N'-bis-2-ethanesulphonic acid) buffer (6.1g of PIPES in 100ml of distilled water, add 16.2ml of N NaOH and make up to 200ml with distilled water) at pH 7.4.
3. Place the specimens in a volume 10–20 times greater than the total volume of the sample and make sure the samples remains fully immersed in the stabilization solution.
4. Stabilize the sample by gentle agitation for 30–60 min at room temperature.
5. Wash the specimens for 30 min in three changes of the same buffer used to stabilize the specimen.
6. Post stabilize the sample for 1–2 hour in an un-buffered aqueous 1% osmium tetroxide (OsO$_4$) solution at room temperature.
7. Wash the sample twice with de-ionized water at room temperature.
8. Further stabilize the sample for 30 min in a 0.5–1% aqueous solution of uranyl acetate (C$_4$H$_6$O$_6$U) at room temperature.
9. Briefly wash the sample in de-ionized water.
10. Ensure the sample remains wet and proceed immediately to dehydration. (see Chapter 6)

Table 8.7. A general schedule for stabilizing plant material by immersion

1. Cut the sample into 1–2mm³ cubes.
2. Prepare a primary stabilization solution of 2.5% glutaraldehyde plus 2% formaldehyde with a 2mM calcium chloride in a 0.1M phosphate buffer (61ml of 0.2M disodium hydrogen phosphate and 39ml of 0.2M monosodium dihydrogen phosphate made up to 200ml with distilled water) at a pH of 7.0.
3. Place the specimens in a volume 10–20 times greater than the total volume of the sample. Make sure the sample are fully immersed in the stabilization liquid. If the tissues float, remove any residual pockets of air by applying a vacuum of no more than half an atmosphere for a few minutes.
4. Stabilize the sample using gentle agitation for 2 hours at room temperature.
5. Wash the specimens for 30 min in two changes of the same 0.1M phosphate buffer pH 7.0 used for the sample primary stabilization.
6. Post stabilize the sample for 1 hour in a 1% osmium tetroxide solution in the same phosphate buffer at room temperature.
7. Wash the sample twice with de-ionized water at room temperature.
8. Further stabilize the sample for 30 min in a 1% aqueous solution of uranyl acetate at room temperature.
9. Briefly wash the sample in de-ionized water.
10. Ensure the sample remains wet and proceed immediately to dehydration.

vacuole membrane, and in photosynthetic organisms, the double chloroplast membrane, are osmotically sensitive. They generally contain less protein than animal cells but frequently contain a wide range of secondary chemical products. Table 8.7 gives a generally satisfactory introductory stabilization procedure that may need adjusting for particular samples.

8.2.5.7 A General Total Immersion Protocol for Vertebrate Animal Material

Unlike plants, the outer region of individual cells is limited only by a thin membrane. The cell contents contain the same organelles found in plant cells with the exception of chloroplasts and large vacuoles. Many of the cells show specialization and are frequently organized into discrete tissue types, which in turn are combined into a variety of larger interconnected and specialized functional organs and tissues. There are no single-cell vertebrates. The exterior of multicellular organisms are bound by a layer of strengthened skin and the whole animal is supported by an internal tough skeleton of articulated bones (Table 8.8).

8.2.5.8 A General Vascular Infusion Protocol for Vertebrate Animal Material

In the case of large specimens, such as most terrestrial and aquatic vertebrates, the stabilization solution may be perfused through the internal vascular tissues of either the whole animal

8.2 Sample Stabilization for Imaging in the SEM Using Chemical Procedures

Table 8.8. A general schedule for stabilizing vertebrate animal material by immersion

1. Cut the sample into 1mm^3 cubes or 0.5mm thick sections.
2. Prepare a primary stabilization solution of 4% glutaraldehyde plus 2% formaldehyde and 2mM calcium chloride in a 0.1M PIPES buffer (6.1g of PIPES to 100ml of distilled water, add 16.2ml of N NaOH and make up to 200ml with distilled water) at pH 7.4.
3. Place the specimens in a volume 10–20 times greater than the total volume of the sample.
4. Keep the tissue gently and continuously bathed in the stabilization fluid at 2–4C° for 1 hour and continually add drops of the cold stabilization fluid.
5. Wash the specimens for 30 min in two changes of the same 0.1M PIPES pH 7.4 used for the primary stabilization fluid.
6. Post stabilize the sample for 1 hour in a 1% aqueous solution of osmium tetroxide solution at room temperature.
7. Wash the sample twice with de-ionized water at room temperature.
8. Further stabilize the sample for 30 min in a 1% aqueous solution of uranyl acetate at room temperature.
9. Ensure the sample remains wet and proceed immediately to dehydration

or isolated organs such as the kidney, lung, or heart. This process can be used to first clean and then stabilize the vascular tissue and then the stabilization fluids slowly allowed to penetrate the associated organ, which either remains attached to the body of the animal or is removed from the body. The perfusion technique, which was originally designed for preparing samples for transmission electron microscopy, is particularly useful for SEM, as most secondary electron images arise from the first 5–10 nm of the vascular surface. The surfaces may be relatively quickly stabilized by most chemicals.

The animal is anesthetized and either a particular whole organ, i.e., the liver, exposed by surgical dissection or the whole body via the heart. Great care must be taken with the surgical procedures to avoid damaging the fragile tissues and distorting their contents. A suitable size canula attached to a large tube, is inserted into an appropriate artery of the organ. For whole body perfusion, the canula is inserted into the arterial system via the left ventricle of the heart. *Note:* These procedures are only permitted in properly licensed laboratories.

The tube is connected to a bottle containing either a balanced cleaning fluid or a stabilization solution, and held approximately 1 m above the specimen, which is placed at a slightly inclined angle. The solution passes into the animal either under constant light gravity flow pressure or pumped into the animal at a constant volume. Suitably solutions are allowed to slowly irrigate either the whole system or a particular organ.

When perfusing an organ, a cut is made in an appropriate vein to allow the fluids to leave the vascular system. With whole body

Table 8.9. Pre-stabilization and stabilization fluids for use with vascular diffusion of small mammals

Pre-stabilization fluid
10 mM PIPES buffer pH 7.4 containing:
140 mM sodium chloride
2.7 mM potassium chloride
2 mM calcium chloride
5 mM sodium nitrate
20 mM glucose
2.5% polyvinylpyrrolidone (MW = 40,000)
Stabilization fluid
100 mM PIPES buffer pH 7.2–7.4 containing:
2–4% glutaraldehyde
0.5–4% formaldehyde
2 mM calcium chloride
2.5% polyvinylpyrrolidone

From Glauert and Lewis, 1998.

perfusion, the fluids leave the vascular system through a cut made into the right auricle. These fluids are constantly mopped up and removed. The perfusion process usually takes 10–15 min at ambient temperature. Care has to be taken to control the pressure of the liquid to avoid internal organ rupture.

The perfusion fluids must have adequate buffering capacity and a suitable osmolarity because of the release of acids when using aldehyde stabilization. Glauert and Lewis (1998) recommend using a 0.1 M cacodylate, phosphate or PIPES buffer at a pH between 7.2–7.4. The perfusion stabilization must also take account of the colloid osmotic pressure and Glauert and Lewis (1998) recommend using either dextrin of polyvinylpyrrolidone (PVP) solutions of between 40,000–80,000 molecular weight. Table 8.9, derived from Glauert and Lewis (1998), gives details of the stabilization fluids.

This process is dramatically shown during liver perfusion, which changes from its normal dark red color to light brown as the blood is washed out of the organ. Details of the perfusion technique are well described in the book by Glauert and Lewis (1998), from which the following general perfusion method is adapted (Table 8.10).

8.2.5.9 A General Stabilizing Protocol for Invertebrate Animal Material

The cells, their internal organelles, and tissue organization are similar to those found in vertebrates. The invertebrates range in size from single-cell aquatic organisms to multicellular terrestrial forms. Unlike the vertebrates, they do not contain an internal skeleton, but the terrestrial forms are frequently invested with a tough, near impermeable, exoskeleton (Table 8.11).

Table 8.10. A standard perfusion technique for small animals

1. The perfusion apparatus should be assembled and filled with warm saline.
2. The animal specimen should be properly anaesthetized and laid on its back in a shallow tray.
3. Appropriate incisions are made to open the thorax and expose the heart.
4. Clean away the tissues around the heart and place a ligature beneath the arch of the aorta ready to tie the perfusion canula into place.
5. Make a stitch through the apex of the heart and tie the thread to the side of the animal so that the heart is under slight tension.
6. Make a small incision into the wall of the left ventricle and quickly insert a metal canula so that it projects 5mm into the aorta.
7. Tie the canula in place and as soon as the perfusing solution is flowing, cut the right atrium.
8. Continue the perfusion until the volume of the perfusing liquid equals the weight of the animal.

From Glauert and Lewis, 1999.

Table 8.11. A general schedule for stabilizing invertebrate animal material by immersion

1. Where necessary, cut the sample into $1mm^3$ cubes or 0.5mm thick sections.
2. Prepare a primary stabilization solution of 3% glutaraldehyde plus 2% formaldehyde and 2mM calcium chloride in a 0.2M phosphate buffer (mix 80ml of 0.2M disodium hydrogen phosphate and 20ml of 0.2M monosodium dihydrogen phosphate) at pH 7.4.
3. Place the specimens in a volume 10–20 times greater than the total volume of the sample.
4. Keep the tissue gently moving in the stabilization fluid at the invertebrates ambient temperature for 1 hour and continually add drops of stabilization fluid.
5. Wash the specimens for 30 min in two changes of the same 0.2M phosphate buffer pH 7.4 used for the primary stabilization fluid.
6. Post stabilize the sample for 1 hour in a 1% aqueous solution of osmium tetroxide solution at room temperature.
7. Wash the sample twice with de-ionized water at room temperature.
8. Further stabilize the sample for 30 min in a 1% aqueous solution of uranyl acetate at room temperature.
9. Ensure the sample remains wet and proceed immediately to dehydration

8.2.5.10 A General Stabilizing Protocol for Microorganisms

Bacteria and viruses are small to very small unicellular organisms. They are bounded by a limiting cell membrane together with a rudimentary cell wall. Unlike the eukaryotic plant and animal cells which have discrete cell bound internal organelles, the prokaryotic bacteria and viruses lack this structural organization. However, this seemingly simpler structural form can perform much of the sophisticated functional biochemistry found in the eukaryotes (Table 8.12).

The specimen stabilization procedures discussed so far are the most popular approaches and simply involve floating, immersing,

Table 8.12. A general schedule for stabilizing microbial material by immersion

1. Collect the cells by centrifugation.
2. Prepare a primary stabilization solution of 3% glutaraldehyde in a 0.1M phosphate buffer ((mix 80ml of 0.2M disodium hydrogen phosphate and 20ml of 0.2M monosodium dihydrogen phosphate) and make up to 200ml with distilled water) at pH 7.4.
3. Place the specimens in a volume 10–20 times greater than the total volume of the sample.
4. Keep the cells gently moving in the stabilization fluid tissue at 2–4C° for 12 hours.
5. Precipitate the cells by centrifugation and wash for 15 min in two changes of the same 0.1M phosphate buffer used for the primary stabilization fluid.
6. Post stabilize the sample for 30 min in a 1% aqueous solution of osmium tetroxide solution at room temperature.
7. Precipitate and wash the cells twice with de-ionized water at room temperature.
8. Further stabilize the sample for 30 min in a 1% aqueous solution of uranyl acetate at room temperature.
9. Ensure the sample remains wet and proceed immediately to dehydration.

or infusing the samples with stabilization liquids. Small specimens also may be exposed to the vapor phase of stabilization chemicals such as osmium tetroxide.

8.2.5.11 The Use of Microwave Assisted Stabilization

There is good evidence developed over the past 35 years that the use of microwave radiation during the stabilization process improves the sample preservation. Microwaves are a form of non-ionizing radiation that increases the diffusion rate of the stabilization chemicals by agitating their molecules. The method is good for impervious samples or when rapid processing is needed. Much emphasis has been placed on creating an even microwave energy fields and eliminating the heating effect induced by the radiation. The main advantage of using brief exposure to microwave radiation is that it speeds up the processes of chemical stabilization and staining. Organic reactions are enhanced by microwave irradiation and will accelerate resin polymerization. This is not a procedure to be carried out in a domestic microwave oven, and the book by Login and Dvorak (1994) has details of some of the specialized equipment and how this procedure may be carried out.

The specialized laboratory microwave tissue processors are available from a number of companies, including Electron Microscopy Supplies (www.emsdiasum.com), Microwave Research and Applications (www.microwaveresearch.com) and Ted Pella Inc. (www.tedpella.com). Figure 8.7 shows the Ted Pella Pelco BioWave® Pro Microwave Tissue Processor.

Figure 8.7. Pelco BioWave® Pro Microwave Tissue Processor (Picture courtesy of Ted Pella Inc., Redding, CA)

More recent studies with specialized equipment and new experimental protocols put more emphasis on the direct effects of the energy of the microwaves. This approach is much less invasive and is proving to be a successful way to stabilize thermally sensitive molecules which are the target of immunocytochemical studies. The book by Giberson and Demertee (2001) gives details of this new approach to microwave stabilization.

8.2.5.12 Anticipating the Next Step in Sample Preparation

It is important to remember that, for biological material, the processes of sample stabilization are invariably followed by staining, dehydration, and then charge elimination. Staining is not usually necessary in order to obtain images in the SEM, but can be used to localize regions of interest for analysis. Details of these problematic procedures are discussed in the next chapter. For successful dehydration it is important that the stabilized sample is not allowed to dry out and for this reason, the wet stabilized sample should proceed directly to one of the different ways to remove water discussed in Chapter 7. If the sample accidentally dries, the delicately preserved structural features will become so grossly distorted that it should be discarde and the whole process restarted.

8.2.6 *Wet and Liquid Samples*

Most, but not all, samples in this category are of biological origin. However, the natural state of paints, glues, polymer suspensions, adhesives, fibers, drug suspensions, food products, mud, cements, etc., all contain varying amounts of either water or some other liquid. Most of these materials need little stabilization other than the use of heavy metal compounds, i.e., osmium tetroxide, to enhance the coefficient of the SE and BSE signal emitted from light element samples followed by some form of dehydration. Constituent parts of living biological organisms such as animal body fluids and plant saps are either wet or liquid. Many food products and most pharmaceutical agents are ingested into living organisms in either a liquid or wet form. There are several ways to stabilize wet and liquid samples.

8.2.6.1 Examine the Specimens Untreated and Avoid Any Attempts at Stabilization

This approach involves using the wet cells and variably pressure scanning electron microscopes discussed in Chapter 7. These methods have been used to study some polymers, naturally moist biological material, foodstuffs, and pharmaceutical agents.

8.2.6.2 Using Ambient Temperature Physical Means

The liquids and solids in the samples can be separated by physical means, such as filtration, centrifugation, or electrophoresis, and the two components examined in isolation. This would be suitable where there are recognizable solid components in the liquid phase and the approach would be suitable, for example, when examining red and white blood cells separately from the plasma and vascular conducting tissue. The disadvantage of this approach is that the liquid and solid components are studied in isolation and lose any important structural inter-relationships.

8.2.6.3 Using Ambient Temperature Chemical Means

The liquids and solids in the samples can be first separated by physical means followed by chemical means. The solid phases could be examined in isolation and attempts made to selectively dissolve or precipitate components in the liquid phase. This approach would be suitable for examining the water and saline phases in marine oil bearing rocks. The disadvantage of this approach is that in addition to the loss of important structural inter-relationships, the use of chemicals may interfere with other parts of the sample.

8.2.6.4 Using Low Temperature Physical Means

The liquids in the sample are converted to a solid state by rapid cooling and examined at low temperature in the microscope. A detailed discussion of this approach is considered here.

8.3 Sample Stabilization for Imaging in the SEM Using Non-chemical Methods at Low Temperatures

8.3.1 Introduction

Sample stabilization at low temperatures provides high quality images and the retention of chemical information about the specimen. In contrast to the previous chemically driven processes, the non-chemical approach is essentially, non-invasive, for all the liquids, principally water, are transformed in situ from the liquid phase to the solid phase.

Low temperature procedures may be used to stabilize specimens from four of the six groups of specimens considered in this book. This method should not be considered solely as a way to stabilize biological samples and their associated products such as foods and drugs. It also can be used to stabilize frozen components of our environment such as ice, snow, and hail.

Cryo-technology is central to the stabilization of aqueous non-biological systems such as paints, suspensions of solid materials, emulsions, solutions, soils, clays, muds, and cements; of any non-aqueous liquid systems such as oils and organic fluids; even of gases, vapors, and volatile materials. In addition, low temperature technology plays an important part in sample preparation of plastics, polymers, and elastomers. The common physical para-meter of these diverse samples is that, provided their temperature can be kept below their melting or vaporization point, they can all be solidified or show a dramatic decrease in flexibility.

Converting a wet, liquid, soft, or flexible material into a solid provides a firm matrix for their stabilization, manipulation, microscopy, and analysis. The cooling processes ensures keeping in place the natural tri-phasic state that exists in some samples, and that dissolved elements, compounds, and molecules remain in situ within the solidified matrix. Sufficiently low temperatures effectively immobilize dynamic and physiological processes. These low temperature processes are, to a first approximation, a chemically non-invasive preparative procedure and at very low temperatures there is a diminution in high energy beam induced radiation damage.

The single disadvantage of the low-temperature approach to preparing wet, liquid, and aqueous specimens is the possible formation of crystalline material during the liquid–solid phase transition. Many organic materials, for example, ethyl alcohol, when cooled below their melting point, form amorphous glasses in which any suspended or dissolved materials remain randomly distributed. In contrast, as hydrated samples are cooled, they generally form highly ordered eutectic mixtures of pure ice crystals and partitioned solids that bear little structural relationship to their original random aqueous organization. However, if the samples are cooled fast enough, this structural disorganization can either be significantly diminished or entirely avoided.

This section of the book focuses on low temperature methods developed to stabilize the natural structural and chemical composition of biological specimens. These same methods also can be used to stabilize hydrated organic samples, hydrated inorganic systems, and liquid organic samples. These discussions emphasize the practices and procedures needed to ensure good specimen stabilization. A detailed explanation of the physiochemical basis and procedures of low temperature light, transmission and scanning electron microscopy, and x-ray microanalysis, may be found in Echlin (1992). A general outline of low temperate sample stabilization follows.

8.3.2 Some Problems Associated with the Formation of Ice

The primary aim is to convert the highly disorganized liquid phases of samples to an amorphous non-crystalline or glassy solid. This is relatively easy to achieve with most non-aqueous specimens. This is much more difficult with aqueous samples because the water is too readily converted to the highly organized crystalline ice we find in snow flakes and frozen peas. These ice crystals severely distorted the non-aqueous parts of soft wet materials and the damage in biological specimens can be so bad that it is not possible to recognize structural features in the frozen material.

There are about 12 polymorphs of ice depending on the pressure and temperature at which they are formed and maintained. The three naturally occurring ice polymorphs that are significant to low temperature microscopy and analysis are hexagonal ice (I_H), cubic ice (I_C), and vitreous ice (I_V). The term amorphous ice (I_A) is also used interchangeably with the term vitreous ice (I_V). The debates continue (see Echlin, 1992 for details) as to whether the two materials are one and the same thing, but this book considers them synonyms. An additional ice polymorph high pressure ice (I_N) can be formed experimentally by rapidly applying a high pressure that allows the water in the sample to undercool before it forms ice. Table 8.13 provides useful information about these four ice polymorphs.

The four polymorphs of ice stay in their different states at atmospheric pressure, provided they are immediately stored in liquid nitrogen at 77 K. If the temperature of the different ice polymorphs is allowed to rise they will change from vitreous/amorphous ice > cubic ice > hexagonal ice > liquid water. This conversion cannot be reversed.

Table 8.13. Physical features and inter-conversions of four ice polymorphs

Ice phase	Pressure (Kbar)	Lower Temperature (K)	Upper Temperature (K)	Transformation at upper temperature
Hexagonal	0	0	273	Melts
Cubic	0	0	183	cubic > hexagonal
Amorphous	0	0	153	amorphous > cubic
Pressure ice	2	0	243	pressure > cubic

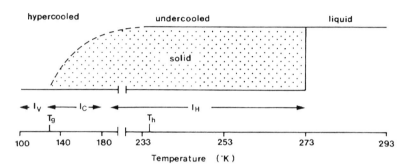

Figure 8.8. The relationship between the forms of stable and metastable water and ice which can exist at subzero temperatures. I_V = vitreous ice, I_C = cubic ice I_H = hexagonal ice (Diagram courtesy of Felix Franks)

8.3.3 How Best to Convert Liquid Water and Aqueous Specimens to Ice

The secret to success is the size of the sample, the speed at which it is cooled, and the sample storage at 77 K once it has been frozen. Vitreous ice can be formed by rapidly cooling (~10^5 K/s) very small quantities of aqueous solutions (≤100 μm^3). Once the vitreous ice is formed, it must be kept below the glass transition temperature (Tg) in order to maintain the random arrangement of the molecules and prevent devitrification and subsequent crystallization. The (Tg) for pure water is 135 K, but in the presence of dissolved solutes, the temperature are a little higher. Figure 8.8 shows the relationship between three different forms of ice and water. If the cooling is fast enough, the liquid water will form vitreous ice. If it is not fast enough, it will form either cubic or hexagonal ice, which cannot be transformed to vitreous ice. The transformation $I_H > I_C > I_V$ will not take place if the temperature is lowered, but the transformation $I_V > I_C > I_H$ all to readily occurs if the temperature is raised.

Using specialist equipment, it is now possibly to routinely rapidly cool very small samples (200 μm^3) and keep them below the I_V glass transition temperature for examination at low temperatures on a cold stage in the electron microscope. On the assumption that the aim of the preparative technique is at best to vitrify the specimen, and at worst to convert the water to microcrystalline ice, the following measures should be taken *before* the sample is quenched cooled.

8.3.4 Physical Manipulation of Specimens Prior to Quench Cooling

1. *Specimen size*. The samples should be as small as possible but large enough to cover the regions of interest. Wherever possible, the surface area to volume ratio of the sample should be maximized. Very small wet solid samples are sometimes difficult to handle and run the risk of drying out before they are quench cooled. As a general rule, samples should no larger than 1 mm^3.
2. *Specimen holders*. These should be as small as possible and made from high thermal conduction materials such as aluminum, copper, silver, and gold. In a recent paper by Hawes et al. (2007), 3.0-mm diameter sapphire discs have been used as specimen holders.

8.3.4.1 Standard SEM Specimen Holders

Samples placed on the large mass of standard SEM specimen holders cool relatively slowly and there is ice crystal damage. The ice crystal damage can be lessened by making sure the sample is on the top of the specimen holder and that it is the first thing to be cooled as it makes contact with the cryogen. Alternatively, the smallest possible pieces of a sample are first quench cooled in a liquid cryogen, and under liquid nitrogen, fixed into a vice-like holder on the pre-cooled specimen holder. The small pieces of material, although cooled rapidly, require delicate manipulation to insert them into the large cold secondary holder(s).

8.3.4.2 Small SEM Specimen Holders

Suitable holders can be made from 1 × 3 mm copper or gold tubes with an internal diameter of 0.6 mm. One end of the tube is slightly tapered and the other end is filled with dental wax, which is then depressed by about 1 mm to form a little cup. Small samples are placed in the holders at room temperature and then ensuring the specimens do not dry, are quickly plunged, specimen face down, into a suitable cryogen. The quench cooled tubes and their samples are rapidly transferred to liquid nitrogen and then the tapered end inserted into pre-drilled tapering holes on larger holders. This is the preferred sample-specimen holder configuration, because it allows very small sample to be cooled rapidly and easily inserted into a larger precooled holder that, in turn, is transferred to the cold stage of the microscope. An example of these small specimen holders can be seen in Figures 4.5 and 4.6 in Chapter 4.

Although one should always aim at making the sample as small as possible, the two types of mentioned specimen holders can be modified to accommodate a wide range of sample sizes. Microorganisms, particles, liquid suspensions, and macromolecules can be suspended either as thin films or as microdroplets on very thin highly conductive metal grids or foils or ceramic supports that are then quench cooled. Tiedemann et al. (1998) showed that it is possible to culture individual cells inside 200-μm cellulose microcapillaries that may them be quench cooled in situ. In all these procedures, care must be taken to ensure the sample remains fully hydrated before they are frozen.

8.3.5 Sample Pre-treatment Before Quench Cooling

8.3.5.1 Maximize Sample Under-Cooling

The sample should be under-cooled as much as possible without altering its natural state and before ice crystals begin to form. The amount of under-cooling that may be achieved in biological material is a compromise between retaining the natural physiological processes and preventing any ice crystals from forming. For example, a piece of mammalian tissue could be slowly cooled to 280 K without any deleterious effects. In non-biological material it is probably possible to carefully under-cool wet and liquid samples to just (1 or 2°C) above their freezing point, just before they are quench cooled.

8.3.5.2 Altering the Nucleation Process by Physical Means

As liquid water is cooled, the random movement of the water molecules begins to slow down and produce clusters of the correct size and shape for the condensation of further water molecules that form the nucleus of an ice crystal. One approach is to apply a high pressure to the sample momentarily before it is rapidly cooled. This approach has the effect of dramatically lowering the temperature at which nucleation will occur. This particular approach is used and embodied in the hyperbaric (high pressure) cooling procedure discussed later in this chapter.

8.3.5.3 Artificially Depressing the Sample Freezing Point by Chemical Means

This may be achieved by using cryoprotectants, a group of polyhydroxy chemicals that suppress ice crystal formation. They are mainly used as antifreeze agents in biological systems and fall into two main classes:

- *Penetrating cryoprotectants*. Samples are infiltrated with solutions of chemicals such as glycerol, methanol, ethanol, ethylene glycol, and some sugars (the most commonly used) and dimethylsulfoxide. Low molecular weight monosaccharides and some disaccharides readily penetrate biological membranes. Other cryoprotectants are quite large molecules, and it is necessary first to permeabilize the cell membranes in order for them to get inside the cells and tissues.

- *Nonpenetrating cryoprotectants.* The chemicals work outside cells and tissues and include such organic materials such polyvinylpyrrolidone (PVP) and hydroxyethyl starches of varying high molecular weights. It is necessary to chemically treat the specimens in order for the large antifreeze agents to penetrate the cells and tissues.

Both types of cryoprotectants are very soluble in water and form strong interactions via their hydrogen bonding potential. They lower the equilibrium freezing point and promote undercooling before ice nucleation is initiated. They also increase the viscosity of the medium, which in turn decreases the mobility of water molecules and slows the rate of ice crystal growth and thus promotes vitrification. Cryoprotectants have proved to be very useful in preparing samples for structural studies in both the TEM and the SEM, even though some of the chemicals are toxic and have their own catalog of artifacts. More details of these procedure may be found in Echlin (1992).

8.3.5.4 Chemical Stabilization

It might be appropriate to consider some of the very mild and precise stabilization procedures discussed earlier in this chapter that serve to cross-link sensitive epitopes to macromolecules for subsequence imaging and analysis.

In summary, care must be taken when using chemicals to maximize sample under-cooling, encourage nucleation, and depress the freezing point of water in biological samples to ensure they do not damage the samples. This is probably less of a problem in wet and liquid non-biological specimens.

8.3.6 *Quench Cooling*

The rapid removal of heat from the sample is the most critical part of the preparation process. This process is commonly referred to as quench cooling and should not be confused with the erroneous term "shock cooling" for samples cooled at a rate of 266 K/min. Quench cooling is now a fairly routine procedure and generally produces frozen specimens that only contain microcrystalline ice. The materials used to quench cool samples are referred to as primary or secondary cryogens. Liquid nitrogen is a popular *primary* cryogen that is used to cool organic gases or metals that are referred to as *secondary* cryogens, all of which should have the following properties:

1. A low melting point and a high boiling point.
2. A high thermal conductivity and thermal capacity.
3. A high density and low viscosity at their melting point.
4. They should be safe, inexpensive, easy to use, and readily available.

8.3.6.1 Liquid Cryogens

Sample rapid cooling takes place through two processes: conductive cooling, which relies on direct contact between the sample and its surroundings, and convective cooling, which relies on the mass transfer of heat by circulation of the cryogen past the surface of the sample. Table 8.14 provides information about some chemicals that may be used as liquid cryogens.

8.3.6.1.1 Liquefied Organic Gases

These are popular and effective secondary cryogens because they remain liquid over a wide temperature range, and their use does not need any elaborate equipment. These are readily available and inexpensive, but inflammable and potentially explosive. Gases such as ethane, iso-pentane, and propane can be easily cooled to their liquid state using liquid nitrogen as the primary cryogen. Camping gas, a 80% butane/20% propane mixture derived from natural gas, is not an effective cryogen because of its unfavorable physiochemical properties. Liquefied organic gases are potentially dangerous because they are below the boiling point of oxygen (90 K) that can condense onto the liquefied gas. For this reason only small amounts of the liquefied gases should be used and the whole process should be confined to a spark-free fume cupboard that can be continually flooded with nitrogen gas.

Table 8.14. Thermal and physical properties of some commonly used liquid and solid cryogens

Cryogen	Melting point (K)	Boiling point (K)	Thermal conductivity (J M^{-1} S^{-1} K^{-1})	Viscosity at melting point (poise)
Ethane	90	184	0.24	9×10^{-3}
Isopentane	113	301	0.18	–
Propane	84	231	0.22	87×10^{-3}
Liquid helium	2.2	4.3	0.02	0.2×10^{-3}
Liquid nitrogen	63	77	0.13	1.5×10^{-3}
Slush nitrogen	63	77	0.13	–
Pure Copper	–	–	570 (at 77K)	–
Sapphire	–	–	960 (at 77K)	–

8.3.6.1.2 Liquid Nitrogen
This is inexpensive, readily available, and non-explosive. It is an excellent primary cryogen but a poor secondary cryogen because the melting point (63 K) and boiling point (77 K) are so close to each other that any object placed into the liquid is quickly surrounded by a layer of poorly conducting nitrogen gas because of surface film boiling.

8.3.6.1.3 Slushy Nitrogen
A mixture of solid and liquid nitrogen is easily formed by boiling liquid nitrogen under reduced pressure. The boiling process lowers the temperature of the liquid and small pieces of solidified nitrogen are formed. There is less film boiling because when a sample is placed in the mixture, it first melts the solid nitrogen before causing the liquid to boil.

8.3.6.2 The Application of Liquid Cryogens

These may be used in three different ways to cool the sample by a combination of conductive and convective cooling.

8.3.6.2.1 Immersion, Impact, or Plunge Cooling
The sample may be rapidly plunged, by hand, into the melting liquid cryogen. Figure 8.9 is a diagram of the essential features of an immersion cooling device that can be easily made in a laboratory with only a modest workshop.

A number of companies make more sophisticated immersion or plunge cooling devices. Electron Microscope Sciences (www.emsdiasum.com) make the EMS-002 Rapid Immersion Freezer, and Leica Microsystems (www.leica-microsystems.com) make the even more sophisticated Leica EM CPC Universal Cryofixation and Cryopreparation System shown in Figure 8.10. Much faster cooling rates may be achieved using a spring-loaded specimen holder to rapidly propel the sample into a long narrow container filled with the secondary cryogen, i.e., propane or ethane, kept just above its melting point by liquid nitrogen. A recent paper on plunge cooling by Ge et al. (2008) gives details of a new controlled environment vitrification system for preparing small wet sample for low temperature SEM.

The size and shape of the sample is important, and it must travel as far as possible through the cryogen. The mean cooling rate is $1\text{--}2 \times 10^4$ s^{-1} K^{-1} and will provide a 5–10 μm deep ice crystal free zone.

8.3.6.2.2 Spray Cooling
This simple device shown in Figure 8.12 involves spraying aerosol droplets of the sample on to a clean, polished copper

8.3 Sample Stabilization for Imaging in the SEM Using Non-chemical Methods

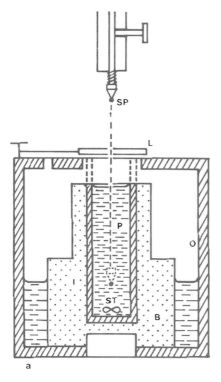

Figure 8.9. A diagram of a simple plunge cooling device. SP = spring loaded specimen holder, L = lid, P = liquid propane, ST = stirrer, B = large aluminum block. The inner container is made of copper. The outer container (O) and the inner containers contain liquid nitrogen

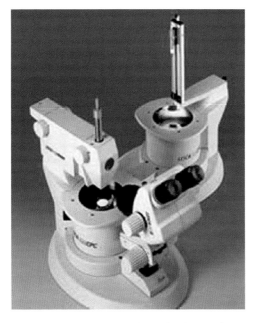

Figure 8.10. The Leica EM CPC combined plunge freezing and metal mirror freezing cryopreparation system (Picture courtesy of Leica Microsysteme GmbH, Vienna, Austria)

8. Sample Stabilization for Imaging in the SEM

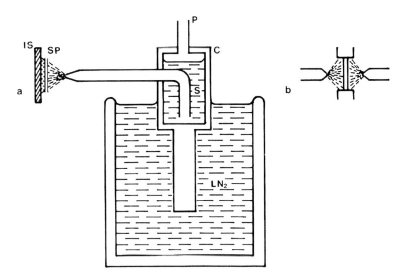

Figure 8.11. Diagram of (A) single and (B) double jet-cooling device. P = pressure device, C = closed chamber, S = secondary cryogen, i.e., propane, SP = specimen suspended on a thin metal foil. LN$_2$ = liquid nitrogen

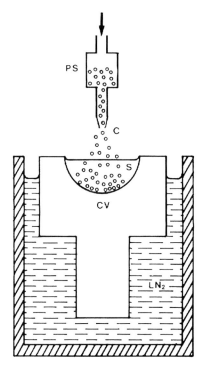

Figure 8.12. A diagram of a spray cooling device. PS = particles suspended in a pressurized capillary, C = microdroplets containing specimens. CV = copper block, S = secondary cryogen, i.e., propane, LN$_2$ = liquid nitrogen

surface, such as a TEM specimen grid, cooled with liquid nitrogen. It is important that the metal surface is not contaminated or covered with liquid nitrogen. This is a very effective way to cool microorganisms and small particles. The equipment can be made easily in the laboratory. The mean cooling rate is 5×10^4 s^{-1} K^{-1} and will give a 20- to 30-µm deep ice crystal free zone.

8.3.6.2.3 Jet Cooling

A diagram of the essential features of this device is shown in Figure 8.11. A jet of liquid cryogen, such as ethane or propane cooled by liquid nitrogen, is squirted under pressure simultaneously onto one or both sides of a suspended specimen help in place on an appropriate sample holder. This is an effective way to quench-cool thin flat specimens, but the equipment is specialized and the JFD 030 Jet Freezing Device is available from Bal-Tec AG Balzers (www.bal-tec.com). The mean cooling rate is 3×10^4 s^{-1} K^{-1} and will give a 10- to 15-µm deep ice crystal free zone.

8.3.6.3 Solid Cryogens

This is a very effective way to rapidly quench cool soft biological and wet samples. Certain materials, such as a highly polished block of very pure copper coated with a thin layer of gold or a block of sapphire, have a very high thermal heat capacity; that is, they have the property of absorbing heat without any increase in temperature. The sapphire blocks are cooled with liquid helium at 17 K and the copper blocks with liquid nitrogen at 77 K. A diagram of the essential features of this device is shown in Figure 8.13. The process

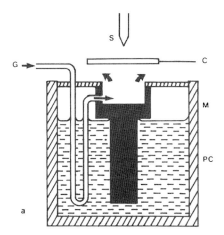

Figure 8.13. Diagram of an impact cooling device. S = spring loaded specimen, G = pre-cooled dry nitrogen or helium gas to prevent frost forming on the metal surface, C = shutter, PC = Dewer filled with liquid nitrogen or helium. The black container is made of copper with a highly polished surface

is referred to as, impact, slam, or metal mirror cooling because the specimen is pressed rapidly against the cold surface, which must be scrupulously clean of contamination and liquid cryogen.

Suitable commercial equipment is the EM-CPC Cryofixation system from Leica Microsystems (www.leica-microsystems.com) was mentioned earlier. It is important that there is immediate and continuous contact between the sample and the cold surface because any specimen bounce significantly reduces the effective cooling rate.

This is a very effective way of cooling the surface layers of quite large thin specimens. The physical contact region of the specimen with the metal surface will show some mechanical deformation but at 1–2 μm below the surface, there is excellent conductive cooling. The mean cooling rate is 2.5×10^4 s^{-1} K^{-1} and gives a 20-μm deep ice crystal free zone. Tables 8.14 and 8.15 give the physical properties and the relative cooling rates of a number of different liquid and solid cryogens.

8.3.6.4 High Pressure Cooling

An examination of a phase diagram of water will show that vitreous water should form when liquid water is rapidly cooled under high pressure. As water freezes it increases in volume, and high pressure will hinder this expansion, and in turn hinder crystallization. This effect is manifest in a decrease in the freezing point, a reduction in the nucleation rate and the rate of ice crystal growth. The freezing point of water at approximately 2 kbar is reduced to 250 K, although under-cooling will continue on down to 183 K. To put it simply, high pressure cooling lowers the solidification point of water by 90°. A diagram of the essential features of the high pressure cooling device is shown in Figure 8.14.

Complex and expensive pieces of equipment are needed to achieve vitrification of small specimens by cooling them with high pressure liquid nitrogen. The high pressure, typically 2–3 kbar, vitrifies some specimens to a depth of 200 μm. Using specialist equipment, it is now possible to routinely rapidly cool very small samples (200 μm^3) and keep them below the I$_V$ glass transition temperature for examination at low temperatures in an electron

Table 8.15. The mean cooling rate and range of commonly used liquid cryogens

Cryogen	Temperature (K)	Mean cooling Rate (10^3 K s^{-1})	Range (10^3 K s^{-1})
Ethane	90	15	10–500
Isopentane	113	9	1–90
Propane	83	12	1–100
Liquid nitrogen	77	0.5	0.03–16
Slush nitrogen	77	2	1–21

8.3 Sample Stabilization for Imaging in the SEM Using Non-chemical Methods 171

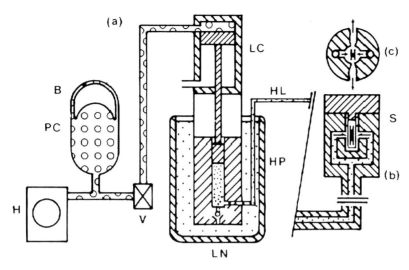

Figure 8.14. Diagram of a high pressure cooling device. H = hydraulic pump, PC = high pressure container, V = valve, LC = low pressure cylinder, HP = high pressure cylinder containing liquid nitrogen. HL = high pressure line, S = metal foil sandwiched sample

Figure 8.15. Leica EM PACT2 portable high pressure freezer (Picture courtesy of Leica microsysteme GmbH, Vienna, Austria)

microscope fitted with a cold stage. Alternatively, the frozen specimens may be processed through the freeze-drying and/or the freeze substitution methods discussed in Chapter 7.

There are two commercially available instruments; the Leica Microsystems, Leica EM PACT High Pressure Freezer (www.leica-microsystems.com) shown in Figure 8.15 and the Bal-Tec HPM 100 High Pressure Freezing Machine (www.bal-tec.com).

Both instruments are capable of producing frozen specimens that are devoid of visible ice crystal artifacts when examined in the TEM. It is more difficult to achieve this type of preservation in the SEM, which usually examines bigger samples that, alas, may only be cooled slowly.

High pressure freezing is providing an important preparative technique for biological specimen preparation and a number of studies have been made to improve the way samples are prepared for freezing. The assembled papers arising from a conference in Germany in 2002 are to be found in the *Journal of Microscopy* (Hohenberg et al., 2003) and from a second conference in Germany in 2006 in a more recent publication by McDonald and Muller-Reichert (2008). These papers are a good source of the continuing developments in high pressure freezing.

8.3.6.5 Summary

There are many different ways to cool samples for low temperature microscopy and in order to design the best method for a particular specimen for SEM, we need to optimize all the stages in the process. There is no single perfect method of quench cooling, and an examination of the literature shows a very wide range of cooling rates. Table 8.16 provides a comparison of the different mean cooling rates for the five methods of sample cooling.

The easiest and least expensive way of quench cooling samples for SEM and x-ray microanalysis is to either plunge small sample into liquid propane or larger samples into slushy (melting) nitrogen. The best results are obtained using high pressure cooling, but the expensive equipment can only accommodate 1-mm^3 sample.

8.3.7 A Quench Cooling Procedure for Small Sample Using Liquid Propane

1. The equipment consists of a 25-mm diameter, 50-mm deep stainless steel container held by a movable suspension over a

Table 8.16. Comparison of the mean cooling rates, the range reported in the literature and the depth of the ice crystal free zone, of the five different methods of sample cooling

Method	Mean cooling (10^3 s^{-1} K^{-1})	Reported range $\times 10^3$	Depth of ice free zone (μm)	Best cryogen (K)
Plunge	10–12	2–500	2–10	Ethane (93)
Spray	50	8–100	20–30	Propane (83)
Jet	30–35	7–90	10–15	Ethane (93)
Impact	25	6–55	15–25	Copper (20)
High Pressure	0.5	–	250	Nitrogen (77)

8.3 Sample Stabilization for Imaging in the SEM Using Non-chemical Methods

1 l Dewer. During operation, the equipment must be kept in a spark free fume cupboard.
2. The Dewer is filled with approximately 750 ml of liquid nitrogen, and once it has stopped bubbling, the empty stainless steel container is lowered approximately 40 mm into the liquid nitrogen and allowed to cool.
3. The metal end of the gas delivery tube connection to the regulator valve on the propane gas container is cooled in the liquid nitrogen until it stops bubbling.
4. The cold metal tube is then placed at the bottom of the now cold stainless steel container and the regulator valve on the gas container is *slowly* opened. The propane quickly cools to a liquid that should be allowed to fill the metal container to about 5 mm from the top.
5. The gas is turned off and a small loose fitting lid is placed over the liquid propane filled container and the liquid is allowed to solidify. It is important to keep checking the levels of liquid nitrogen. The quench cooling device is now ready to use.
6. As has already been shown, there are a large number of different specimen holders that can be used, but for damp, liquid, and wet specimens, one of the best specimen holders can be made as follows.
7. A 3-mm long containers can be easily made from 1-mm thick, internal diameter 600-μm high purity silver tubes. One end of the tube is sealed with a small piece of dental wax in which a 1- to 2-mm depression is made. It is best to prepare a dozen or so of these holders before they are loaded and quench frozen.
8. A small warm metal rod is placed into the center of the now solid cryogen to create a pool of melting propane liquid surrounded by solid propane.
9. The open end of the silver specimen holder is filled with a small amount of the specimen using either a fine pipette or a toothpick. In order to prevent any incipient drying, the specimen should only be placed into the end of the silver tube *just before* they are to be quench cooled.
10. Using a pair of insulated fine forceps, the small specimen holder with the sample at the leading end is plunged into the melting cryogen and held in place until any boiling has ceased. While still holding onto the specimen holder it should be drawn up from the liquid propane, gently taped against the side of the metal container to remove any remaining liquid propane and then dropped into the surrounding liquid nitrogen. This sequence of events needs split second timing and should take no more than 2–3 s!
11. The frozen specimen must remain under liquid nitrogen and either placed in a pre-labeled vial and stored in a liquid nitrogen refrigerator for later use, or transferred to a larger specimen holder while still under liquid nitrogen, quickly moved

to the pre-cooled cold stage of the SEM. The frozen specimen should not be allowed to warm up.
12. Once the quench cooling is finished, the liquid propane–liquid nitrogen metal container should be allowed to slowly evaporate in a Class 1 spark free fume cupboard.
13. It is important to remember the following safety features regarding the use of propane and other organic gases:
   a. Liquid organic gases are heavier than air.
   b. Oxygen from the air will condense into the cold liquids.
   c. The gas alone is very inflammable.
   d. The gas and oxygen mixture is potentially explosive.

8.3.8 Post Cooling Sample Processing

Once the sample has been successfully quench cooled and stored in liquid nitrogen, there are a number of different ways to handle the cold specimen in such a way that they do not melt or become covered in condensed water vapor.

8.3.8.1 Transfer to Another Piece of Equipment

1. They may be stored at low temperatures in liquid nitrogen.
2. Their internal contents can be exposed.
3. They can be immediately examined on a low temperature stage in the SEM.
4. The water can be removed either by physical or chemical means.

The one central feature of these four options is that the sample must be kept, at all times, well below the glass transition temperate Tg, which for pure water is 135 K, in order to prevent de-vitrification and subsequent crystallization. The cold samples are usually very small and any exposure to air will very quickly initiate the de-vitrification, melting and drying processes that will destroy the specimen.

8.3.8.1.1 Low Temperature Sample Transfer
Specimens should not be exposed to air even at low temperatures because of the high probability of water vapor condensation. If necessary, the samples can be washed in clean liquid nitrogen filtered through cotton gauze to remove minute ice crystals. It is most important that the frozen sample does not undergo re-crystallization or ice contamination and all transfer between the quench cooling device should, ideally, be under liquid nitrogen or in an assured environment of dry, cold nitrogen gas, and/or in vacuum.

The easiest and most reliable way to transfer frozen samples is to make sure they are always handled with insulated cold tools and always remain immersed in liquid nitrogen. The simplest and most reliable transfer containers can be cut to size from thin walled plastic bottles. In order to diminish the amount of liquid nitrogen bubbling, the plastic containers should be placed in tight fitting holes cut from pieces of polystyrene foam. Sometimes, the samples and their specimen holders are so small that they can only be reliably viewed using a binocular microscope.

The most convenient way to view and manipulate these small samples is to place them in a glass Petri dish sitting in an outer larger Petri dish both of which are filled with clean liquid nitrogen. This simple arrangement ensures there is no bubbling in the inner dish. The transfer devices need to be made to measure in order to fit into the equipment, procedures and spatial location of the low temperatures facilitate in a given laboratory.

8.3.8.1.2 Low Temperature Storage
The quench cooled samples should be quickly removed from the cooling device under liquid nitrogen using liquid nitrogen–cooled tools and placed into small plastic vials filled with liquid nitrogen. These vials should be properly labeled and inserted in the metal slots of the cane-like rods that are a standard feature of liquid nitrogen refrigerators.

8.3.8.1.3 Exposing the Internal Contents of the Frozen Specimen
The general procedures for sample exposure were considered in Chapter 6, but the specialized techniques dealing with frozen sample are discussed in the following.

In some cases, it is only the surface features of the sample that are of interest. Provided this surface is undamaged, uncontaminated and the sample is kept below 123 K in a clean, high-vacuum environment, the specimens may be imaged directly following transferring them to the cold stage of the SEM. More often, it is the inside of the sample that is of interest and it is appropriate to consider the three ways it is possible to expose the inside of a frozen specimen. The mechanical processes involved in these activities are no different from the methods discussed earlier in Chapter 6. However, there are additional complications of having to do this at low temperature.

8.3.8.1.4 Cryosectioning
Cryosectioning is a flow process involving extensive deformation of the ice as the material is thinly sliced at the tool (knife) and sample interface. The big advantage of any sectioning process is that the interior of the specimen can be revealed as a flat region at a pre-determined place of the sample. The resulting sections can be viewed in the fully frozen hydrated state or after the ice has been sublimed by freeze drying. There are different categories of sections, depending on their thickness. Ultrathin sections

are 50–100 nm thick and usually only examined in the TEM, and thin sections are 1–5 μm, which are suitable for examination in the SEM. The Leica EM-1900 cryomicrotome from Leica Microsystems (www.leica-microsystems.com) can be used to cut sections between 1- to 60-μm thick.

Cryoultramicrotomes are specialized pieces of equipment for cutting frozen samples and operate at between 123 and 193 K. Suitable instruments are the CR-X Cryosectioning Attachment from Bal-Tech AG, Balzers (www.bal-tec.com) and the Leica EM FC6 from Leica Microsystems (www.leica-microsystems.com). Knives are made from glass, diamond, sapphire, or metal. Specialized design cryo-diamond knives are best for cutting thin, 0.25-μm sections and tungsten carbon knives can be used to cut thicker 1-μm sections.

The actual procedures for cryo-sectioning are not discussed here because they vary depending on the type of equipment used. It is most important to follow the operating instruction of the particular model of the cryomicrotome. It is useful, however, to consider certain general operational procedures. The frozen specimen should either be mounted mechanically on the microtome sample holder or stuck in position using a suitable cryoglue such as melting toluene. Alternatively, the thin-walled silver tube in which the sample was initially quench frozen can be inserted into a hole on the microtome sample holder. Wherever possible, the sample mounting should be done under liquid nitrogen. The microtome sample holder and its securely fixed specimen should be transported, under liquid nitrogen, to the pre-cooled working chamber of the cryo-microtome, fixed to the pre-cooled specimen cutting arm, and allowed to equilibrate to the working temperature of the equipment.

The operational temperature should be low, ideally 133 K, the sectioning thickness controls set to give the thinnest sections, and the cutting speed set to the lowest value. If it is not possible to cut continuous sections under these conditions, the temperature, cutting speed, and section thickness should be slowly increased until continuous, smooth sections are obtained. These values vary very much depending on the chosen material and the success of the initial quench cooling process.

As a general rule, samples that are naturally soft and have been successfully vitrified or contain microcrystalline ice appear transparent. Such material can be sectioned more thinly, more easily, and at lower temperatures. Samples containing large ice crystals appear milky, and smooth sections can only be obtained at an increased section thickness and at higher temperatures. At low cutting speeds and higher temperatures (233 K) there is no transient melting of the ice. Thick sections are best cut at 233–253 K, thin section at 193–213 K, and ultrathin sections at 123–143 K.

It is important that the cut sections remain frozen and not allowed to melt or dry. The different types of cryo-microtomes have ingenious devises to collect the frozen sections and transfer them to microscope specimen holders without losing their fully frozen hydrated state. Cryo-sectioning is not always the best way to reveal the interior of a specimen using the transmission mode of the SEM. It is sometimes more convenient to discard the thin sections and examine the smooth, frozen block face using the reflected modes of the SEM such as the SE, BSE, and x-ray photons imaging procedures.

8.3.8.1.5 Cryofracturing
Cryo-fracturing involves the separation of the frozen specimen, by force, along a line of least resistance parallel to the line of the applied force. A fracture face suitable for SEM may be obtained in two different ways. A stress fracture is produced when a sharp blade is used to initiate the fracture and follows through to force the two parts of the frozen sample to separate. A percussive fracture is produced when a blunt tool is hit against the side of a frozen specimen with sufficient force to cause a fracture at some place away and below from the initial contact. The salient features of the two processes follow.

To make a percussive fracture, the lower part of the frozen specimen should be held firmly, with a liquid nitrogen cooled tool, in a vertical direction and maintained well below its glass transition temperature T_g. The blunt tool used to initiate the fracture is also kept at low temperature and then swung horizontally across and toward the top of the frozen sample with sufficient force to snap off the top of the specimen. Care must be taken that the blunt tool does not make contact with the lower part of the frozen specimen. There is usually only a single opportunity to make a percussive fracture.

The same general procedure is used to make a stress fracture. A cold sharp tool, such as a razor blade, is swung horizontally across toward the frozen sample in such a way that as soon as the blade makes contact, its trajectory continues on in a rising horizontal direction above the now fracturing sample. This maneuver prevents the fracturing tool from touching the freshly fracture sample. This type of fracturing allows further fracturing to take place if the first fracture is deemed unsatisfactory.

Both types of fracturing may be simply carried out under liquid nitrogen. The quench cooled sample is held under liquid nitrogen and continually flexed in one direction until it breaks. It is important to remove all debris from the freshly exposed fracture by washing the fracture in clean liquid nitrogen. The disadvantage of this fracturing process is that the essentially random process of fracturing is virtually uncontrollable and it may be difficult to retrieve the fractured sample. Alternatively, the fracture process may be carried out in more specialized equipment that has control over the fracturing process and the environment under which it is performed.

The more complicated cryo-fracturing devices are separate devices, fitted with two air locks, which are bolted to the side of an SEM which, in turn, is fitted with its own cold stage. Figure 8.16 is a schematic diagram of the features of this type of equipment.

Quorum Technologies Ltd. (www.quorumtech.com) produce the CryoSEM System PP-2000T and Emitech Products Inc. (www.empdirect.com) produce the Emitech K-1250 Cryogen Preparation System. Both systems operate at a high clean vacuum and very low temperatures and enable pre-quenched cooled systems to be cryofractured and then treated in different ways. Figures 8.17 and 8.18 are of a modern cryofracturing system attached to an SEM.

The dedicated cryofracturing devices may used as follows:

1. A small (1- to 3-mm^3) specimen is loaded onto a cryofracturing stage transfer device and rapidly quench cooled in slushy

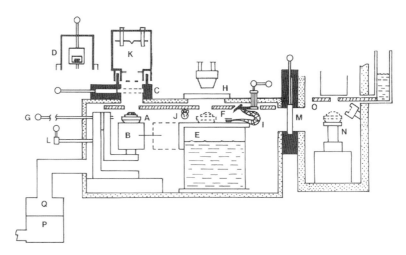

Figure 8.16. Diagram of an integrated low temperature preparation device and low temperature SEM cold stage. The sample is loaded onto a shuttle (*A*) sitting on a pre-cooled metal cube (*B*) via the top high-vacuum gate valve (*C*) using the sealed specimen transfer device (*D*). The shuttle is moved sideways using the shuttle transfer rod (*G*) to the top of the liquid nitrogen tank (*E*) below the cold shroud (*F*). The sample may be observed through the top window and fractured with a microtome knife (*I*) and if necessary, etched with the radiant heater (*J*). A suitable sample may be moved back on the cold cube to the original entry position over which a specimen coating device has been placed. The gate valve is opened and the specimen is coated with a thin layer of a conductive metal. The still frozen fracture-etched-coated specimen on the shuttle is then moved via the second gate valve (*M*) onto the pre-cooled cold stage (*N*) of the SEM for examination and analysis (Picture courtesy of the AMRAY Company)

8.3 Sample Stabilization for Imaging in the SEM Using Non-chemical Methods 179

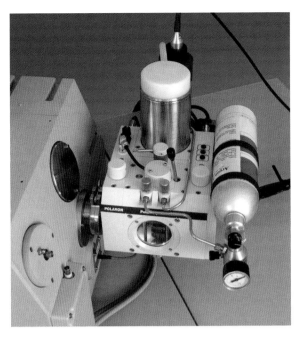

Figure 8.17. The Quorum technologies PP2000 integrated low temperature preparation and SEM cold stage (Picture courtesy of Quorum Technologies, Ringmer, Sussex, UK)

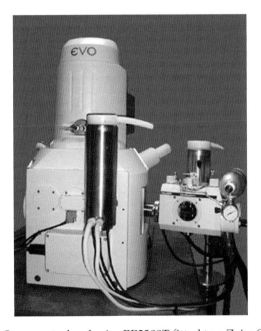

Figure 8.18. Quorum technologies PP2200T fitted to a Zeiss SMT EVO 50 scanning electron microscope (Picture courtesy of Quorum Technology, Ringmer, Sussex, UK)

nitrogen. Alternatively, small samples can be quench cooled separately in one of the melting organic gases and then immediately transferred to liquid nitrogen. The frozen sample is then secured onto the pre-cooled transfer device of the cryofracturing system under liquid nitrogen.
2. The cold, loaded transfer device is quickly moved, via an airlock, onto the pre-cooled cold stage of the fracturing system that is at a controlled low temperature and a very clean high vacuum.
3. The samples, which are kept very cold and held firmly in place, may be repeatedly stress fractured using a cold microtome-driven knife that moves across the frozen sample at an adjustable height. Any frozen debris removed by the knife may be cleaned away and deposited in a very low temperature container help below the fracturing region in order to avoid any subsequent sample surface contamination from subliming ice. The whole fracturing process may be observed through a binocular microscope. The fractured sample can then be treated in three different ways before it is transferred via a second airlock to the pre-cooled cold stage of the microscope:

   a. The fractured sample may be immediately examined.
   b. The sample temperature may be raised to allow controlled etching of the surface ice by sublime.
   c. The surface of the etched or un-etched sample may, if appropriate, be sputter coated with a thin layer of conductive material.

These complicated devices are expensive but are very easy to use and give excellent fracture faces for both image examination and analysis in the SEM.

8.3.8.1.6 Cryoplaning
This procedure provides very flat surfaces of frozen material suitable for x-ray microanalysis and backscattered imaging. Polished surfaces can be produced by two different methods.

1. The quench-frozen sample is cryosectioned at as low a temperature as possible and the sections discarded. The remaining block face will have a very smooth surface.
2. The frozen sample is cryofractured close to the region of interest and the undulating fracture face is polished with fine alumina powder on a metal plate cooled with liquid nitrogen. Care has to be taken in transferring the polished, frozen surface to the SEM and that it remains uncontaminated with either ice crystals or the polishing powder. The one disadvantage of this procedure is that it is very difficult, if not impossible to remove the remains of the abrasive alumina powder.

8.3.8.2 Anticipating the Next Step in Sample Preparation

After exposing the inside of a carefully frozen sample, it is necessary to deal with the frozen liquids within the sample.

8.3.8.2.1 Procedures for Aqueous Samples
There are three possible courses of action to use with wet samples:

1. It is possible to *retain* the water as ice and examine and analyze the so-called frozen hydrated samples at temperatures below Tg.
2. It is possible to *sublime* or freeze dry the frozen water at low temperatures and high vacuum and study the so-called frozen dried sample at either ambient or low temperatures.
3. The third option is to dissolve the frozen ice in organic liquids by a process referred to as freeze *substitution* and then, at elevated temperature, embed the substituted material in a liquid resin which is then polymerized and sectioned.

The processes of freeze drying and freeze substitution are discussed in more detail in Chapter 7.

8.3.8.3 Procedures for Non-aqueous Wet Samples

The preparative procedures for non-aqueous liquids are usually less demanding than those needed for aqueous specimens. Although many organic liquids form a crystalline phase when cooled, the phase change is less disrupted than the changes that occur as water forms ice. If the organic liquid is known to undergo a disruptive phase change as it is cooled, then it will have to be cooled as fast as possible by one or more of the procedures mentioned earlier.

It is usually sufficient to use liquid of slushy nitrogen as the primary cryogen, after which the sample can be fractured or sectioned at low temperatures. Freeze drying and freeze substitution are not recommended for non-aqueous samples. Do not attempt to slowly cool non-aqueous sample on the cold stage of the microscope because this may compromise the vacuum environment of the column. Figures 8.19 to 8.21 are examples of the types of non-biological specimens that may be studied using low temperature SEM.

8.3.8.4 Frozen-Hydrated and Frozen Specimens

If the sample is to be examined in the fully frozen hydrated state, no further preparation is needed. Care has to be taken that the now solidified liquid material remains below its glass transition temperature and uncontaminated during transfer and examination

Figure 8.19. Low-temperature secondary electron image at 120 K of wax particles deposited from diesel fuel. The wax can only be seen at low temperatures as the diesel fuel is volatile at ambient temperatures Marker = 20 mm (Goldstein et al., 2004)

Figure 8.20. Backscattered image of a froze-hydrated fracture face of a core sample of North Sea sandstone saturated with oil. The oil phase is black, the brine appears mid-gray, and the parent rock appears light gray. Image recorded at 93 K. Magnification bar = 100 um (Goldstein et al., 2004)

in the SEM. Most transfer devices have a vacuum interface with the SEM. The fully frozen hydrated state is the best way to examine liquid and soft wet bulk samples, provided they have been quench cooled adequately.

Samples that have not been exposed to chemicals during preparation are probably the closest to their natural state. The same is

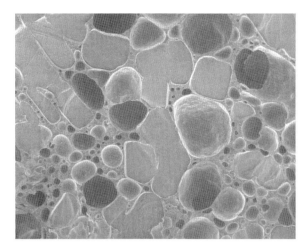

Figure 8.21. An SE image of a fractured face of a sample of fully frozen hydrated ice cream. at 130 K. i = ice crystals, f = fat globules, g = gas bubbles, m = protein–carbohydrate matrix, Image width = 300 mm (Picture courtesy of Mark Kirkland, Unilever Research Centre, Bedford, UK)

true for fracture faces and sections, except that problems of beam damage make it impossible to examine very thin frozen sections in the SEM. Provided there is no ice crystal damage, the quality of the secondary electron images of frozen intact specimens appear the same as specimens prepared at ambient temperatures.

9
Sample Stabilization to Preserve Chemical Identity

9.1 Introduction

In this chapter the process of sample stabilization is more complex. Not only do the procedures have to retain structural integrity, but they also have to retain and localize their chemical components for subsequent analysis. In addition, the stabilization procedures may also have to facilitate recognition of some chemical component. These four requirements, *retention*, *recognition*, *chemical analysis*, and *stabilization* are conjoined and should always be considered together.

Before discussing these procedures it is appropriate to briefly recall the general differences between the topographic (image) contrast mechanisms and the compositional (atomic number) contrast mechanism used in SEM and x-ray microanalysis.

- *Topographic information* (structural images) is provided by the secondary electrons (SE) and the backscattered electrons (BSE) signals.
- *Compositional information* (chemical analysis) is provided by the BSE and x-ray photons (XRP). The physical parameters and operation of these matters are considered at great length in several chapters in the book by Goldstein et al. (2004).

9.2 Operational Parameters Used to Chemically Analyze Samples in the Scanning Electron Microscope

It is important to appreciate the difference between the secondary and backscattered electrons and x-ray photons as these three signals may place certain constraints on the combined topographic and compositional information that can be obtained from the samples:

1. Depending on the accelerating voltage of the primary beam, the BSE and XRP signals are generated deep (μms) below the surface of the sample, in contrast to the shallow (nms) generation depth of the SE signals.
2. The BSE topographic images are generally inferior to the high quality images provided by the SE.
3. The ideal specimen surface for the most accurate and precise analytical information is flat and highly polished in order to reduce surface roughness to no more than 100 nm. This requirement is particularly important for x-ray microanalysis. Resolution of these problems is dependent on sample preparation.
4. The deeper generation of the BSE and x-ray photons in the sample may create operational problems with the analysis of thin films on substrates of particles, rough surfaces, and beam sensitive specimens.

It is convenient to divide the analytical procedure into two sections.

Signals that can provide chemical information about the sample *solely* as a consequence of the interaction of the primary beam with the specimen without prior supplementary invasive chemical treatment of the sample to obtain the analytical information.

Some signals can provide chemical information only *after* invasive specimen modification and stabilization.

Before discussing the sample stabilization procedures in more detail, it is useful to first consider the different ways we can obtain chemical data in the SEM and the general type of information the SE, BSE, and XRP may provide. Table 9.1 lists these signals and their broad operational parameters.

The two main variable parameters of the incoming primary beam are the acceleration voltage and beam current. In simple terms, the higher the voltage the faster the electrons move and the further they penetrate into the specimen. The higher the

Table 9.1. Comparison of the different signals and the range of operational voltage and current commonly used in scanning electron microscopy and x-ray microanalysis. Kv = voltage, A = beam current, SE = secondary electrons, BSE = backscattered electrons, EBSD = electron backscattered diffraction, EDS = energy dispersive x-ray photon spectroscopy, WDS = wavelength dispersive x-ray photon spectroscopy, CL = cathodoluminescence

Signal	Operational voltage Kv	Operational current A
SE	10eV to 30Kv	10pA to 200nA
BSE	500eV to 30Kv	100pA to 200nA
EBSD	10Kv to 30Kv	1–10nA
EDS	1Kv to 20Kv	250pA to 200nA
WDS	1Kv to 20kv	10nA to 200nA
CL	2Kv to 20Kv	300pA to 1nA

beam current, the greater the number of electrons. High accelerating voltage is associated with increased resolution, and high beam current is associated with an increase in the signal emitted from the specimen. The actual voltage and beam current used to examine and analyze a given specimen is very variable and strongly influenced by the following sample parameters:

1. The magnification and spatial resolution needed to obtain the appropriate information from the specimen
2. The type of signals needed from the sample
3. The voltage and current needed to get this information
4. The density and conductivity of the sample
5. The sensitivity of the sample to radiation damage

The broad operating parameters given in Table 9.1 are a measure of the range of accelerating voltage and beam current used to obtain the appropriate analytical signal from the sample. For example, 5 kV and 200 pA enable quantitative analysis to be carried out on a frozen hydrated tea leaf for Mg, Al and Si, whereas 30 kV and 75 nA are used to analyze a super-alloy.

9.3 Methods Needing No Chemicals Added to the Specimen Prior to Analysis

9.3.1 Secondary Electron Imaging: Only of Very Limited and Specific Use in Analysis

The Everhart-Thornley (E-T) detector is used to collect the SE signal from the specimen and provides excellent images in the SEM. The SE coefficient (δ) from pure element targets is relatively insensitive to atomic number. About the only exception is between carbon and gold, which at 20 kV have a nominal value of 0.05 for carbon and 0.2 for gold. This difference makes it possible to image gold particle on a flat carbon substrate as shown in Figure 9.1, but it does not permit the chemical identification of either gold or carbon. We will return to the analytical use of this anomaly latter in this chapter.

9.3.2 Backscattered Electron Imaging: Useful for Distinguishing Differences Between Broad Groups of Elements

Unlike the SE signal, the BSE coefficient (η) increases nearly monotonically with the atomic number of the specimen. For example, ten times more of the incident electron beam is backscattered by gold than by carbon. The differences in the BSE coefficient form the basis of the qualitative analytical procedure. A high-atomic-number inclusion in low-atomic-number material gives a strong BSE signal, which can be used to give sufficient differential contrast in an image, but can only *locate* the inclusion

9. Sample Stabilization to Preserve Chemical Identity

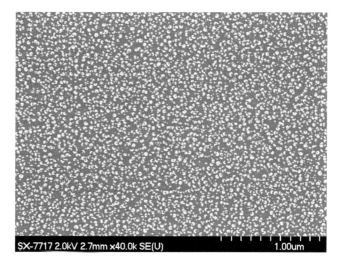

Figure 9.1. Secondary electron image of gold particles on a flat carbon substrate using an E-T collector. Picture courtesy David Joy, University of Tennessee, USA

but not chemically identify it. Backscattered detectors exist in two general forms:

1. The Everhart-Thornley (E T) detector that is most commonly used in the SEM is a combined secondary/backscattered detector. If the detector is given a negative bias, only the higher energy backscattered electrons are detected. Only the backscattered electrons that leave the specimen with trajectories directly toward the face of the detector are collected, all other backscattered electron emitted from the specimen are not collected. The E-T detector usually views the flat specimen at an angle of 30° and gives a geometric efficiency of only 0.8%.
2. A solid state diode backscattered detector forms a thin flat wafer placed on the pole piece of the final lens that has a hole in the middle to allow the primary electron beam to pass through to the specimen below. This type of detector has a large solid angle and collects many more backscattered electrons than the E-T detector described in the preceding.

A full description of backscattered detector is given in Goldstein et al. (2004), and some more recent advantages are in the papers by Erlandsen et al. (2003) and Wandrol (2007).

The BSE signal cannot distinguish differences between individual atomic numbers, only between groups of low and high atomic number materials. For example, the BSE signal could show a heavy metal dust particle inhaled into pulmonary tissue, a metal contaminant on an otherwise clean semiconductor, or a high element inclusion in a polymer. The BSE signal usually

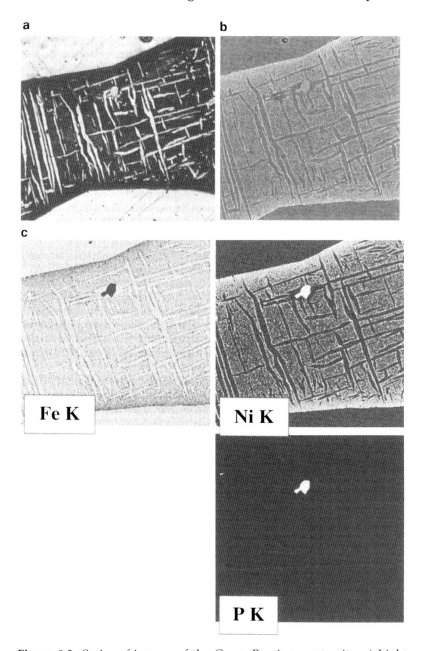

Figure 9.2. Series of images of the Goose Bay iron meteorite. a) Light microscope image of an etched surface, b) BSE image of a non-etched flat polished region, c) BSE images showing the distribution of the Fe, Ni and P regions. The field of view of each image is 0.5 mm × 0.5 mm. Picture courtesy Dale Newbury, NIST, Gathersburg, Maryland, USA

appears bright against a darker background, but it is easy to reverse the contrast in the SEM so that the strong BSE signal appears dark against a light background. Figure 9.2 shows an example of this type of image.

There are four conditions that should be met in order to make full and proper use of this analytical technique:

1. The best signals come from flat, highly polished, conductive samples with no preparation induced surface deformations.
2. There should be maximal specific localization of high-atomic-number elements within the sample.
3. There should be minimal nonspecific distribution of high-atomic-number elements within the sample.
4. The BSE signal must provide an adequate structural images.

Modern BSE annular detectors are simple segmented thin silicon surface barrier diodes with electrical leads that can be mounted just below the final lens concentric to the beam and looking down on the sample surface. By adding or subtracting the signals from one or more of the quadrant of the semiconductor disc, it is possible to separate the BSE signal, to a first approximation, into its compositional and topographic images. Adding the signals enhances the compositional image and enhances the atomic number contrast, whereas subtracting them enhances the topographic contrast.

The BSE imaging procedure has a number of distinct advantages and disadvantages. As shown later, it can be applied to a wide range of different types of specimens without any need of additional preparation. Depending on the acceleration voltage of the SEM, the BSE signal may come from deep within the specimen. Even at low accelerating voltages, the spatial analytical resolution is never high. The resolution also varies with the type of specimen. Thus higher spatial resolution is seen where the compositional contrast predominates in, for example, a finely polished geological sample composed of several distinct phases, a plastic film containing a single heavy metal or a dry organic material, such as a timber section, impregnated with a copper preservative.

The analytical resolution drops away in a fractured metal alloy or the surface of freshly broken rocks because topographic resolution predominates. Although it is not possible to differentiate between two elements that are close to each other, the BSE signal, simply and quickly shows that there may be elemental differences in the sample. The x-ray spectrum can tell precisely the elemental composition of the sample.

9.3.3 Electron Backscattered Diffraction: Crystallography and Phase Information

Electron backscattered diffraction (EBSD) is performed in an SEM by tilting a highly polished bulk specimen at an angle of up to 70° with respect to a 10–40 Kv stationary electron beam. The method is used for two main purposes in the SEM.

It can be used to determine the orientation of known materials. Randle and Engler (2000) measured the individual orientation in a group of small crystals that make up the basic crystallographic units in a specimen. Secondly, as Michael (2000) has shown, by using a combination of crystallography and chemistry it is possible to identify the very small (μm–nm) crystalline phases by determining characteristic crystallographic parameters such as crystalline plane spacing, angles between planes, and the symmetry of crystals in elements. A recent issue of the *Journal of Microscopy* (2008) contains a series of informative papers on EBSD arising from a meeting on the subject in Scotland last year.

The use of EBSD provides quantitative information about crystal orientation, grain boundaries, and phase differences in heterogeneous systems. The spatial resolution of EBSD is related directly using an electron probe size better than the resolution that can be obtained by normal backscattered imaging. The best signals for analysis come from flat surfaces with minimum relief and no preparation induced surface deformation. Figure 9.3 shows and example of an EBSD image.

9.3.4 X-ray Spectroscopy: Small Groups of Elements

X-ray microanalysis alone makes little distinction between the binding states of elements, whereas the chemical stabilization methods are designed around the binding state of the element. For example, consider the binding state of the elements in the three following samples: sulfur covalently bound in a protein,

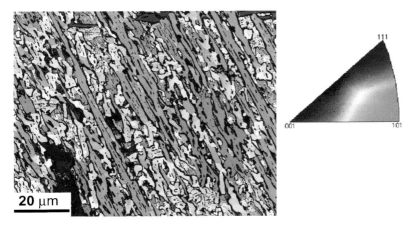

Figure 9.3. An electron backscattered diffraction orientation map of the ferrite (BCC Iron) phase of the Tawallih Valley meteorite. The colour key refers to different crystallographic directions. The uncoated sample has been carefully polished with a final polishing in colloidal silica. Data collected at 20 keV and 10 pa beam current. Picture courtesy Joe Michael, Sandia National Laboratory Albuquerque, New Mexico, USA

calcium in a crystal of apatite, and potassium ions in an aqueous solution. The same element may have different binding states and hence different solubility. The element phosphorus is present as an integrated part of the chemical structure of some body fluids, enzymes, and membranes, whereas the same element is part of the crystalline structure of bone apatite.

X-ray microanalysis can be carried out in three different ways.

9.3.4.1 Energy Dispersive and Wavelength Dispersive Spectrometers

An incident static electron beam is focused on a given point on the sample for a given period of time. The volume of the sample that generates x-ray photons depends on its average atomic number and the voltage and current of the incident electron beam. The emitted x-ray photons can be measured by both their energy, using an energy dispersive spectrometer (EDS) and their wavelength, using a wavelength dispersive spectrometer (WDS). More x-ray photons are generated the higher the voltage and current of the primary beam and the higher the atomic number of the specimen. The whole analytical process is repeated as the static electron beam is moved to another point on the sample.

9.3.4.2 X-ray Mapping

The x-ray signals derived from either of the two types of spectrometer can be used to prepare scanning images, or maps, that contain element-specific information. The detectors are set to collect, either a specific x-ray photon using a WDS or a number of different x-ray photons using an EDS as the primary electron beam scans the whole surface of the sample. The emitted x-ray photons are set to trigger the photo-CRT of the SEM to record a series of white dots that indicate the location of the x-ray photon(s). A secondary electron image can be taken to confirm the location of the x-ray emission. The recent paper by Friel and Lyman (2006) provides an excellent review of the different ways x-ray mapping may be carried out in electron beam instruments. Figure 9.4 shows a typical x-ray map.

9.3.4.3 The General Advantages and Disadvantages of WDS: EDS and X-ray Mapping

9.3.4.3.1 Wds
WDS frequently requires using a higher beam current than EDS; takes a shorter time than EDS to obtain single element qualitative information; takes a longer time than EDS to obtain multi-element quantitative information; and is superior to EDS in obtaining quantitative information.

AutoPhase & X-ray Intensities

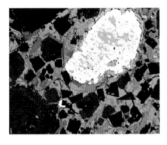
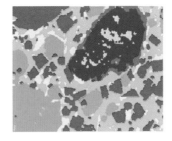

Phase	Na$_2$O	Al$_2$O$_3$	SiO$_2$	P$_2$O$_5$	K$_2$O	CaO	FeO	SrO	Area%
Glass	7.3	3.2	56.5	1.8	8.4	5.7	7.4	0.0	27.5
Combeite	7.9	0.3	66.5	0.8	0.5	23.3	0.7	0.0	23.2
Nepheline	10.3	27.8	51.9	0.5	6.2	0.9	2.0	0.4	19.4
Apatite	0.4	0.2	1.3	43.5	0.2	43.7	0.2	10.4	15.8
Sodalite	20.7	30.1	35.7	0.2	4.2	8.3	0.4	0.0	8.1

Figure 9.4. A highly polished sample of a piece of fine grained igneous rock from the volcano Oldoinyo Lengai in Tanzanier. The left hand picture is a BSE image and the right hand picture is an x-ray map. The chemical analysis of the five phases are given below the two images whose picture width is 400 μm. Picture courtesy, John Friel, Temple University, Philadelphia, USA

9.3.4.3.2 Eds

EDS usually requires a lower bean current than WDS, and takes a shorted time than WDS to obtain multi-element qualitative information.

9.3.4.3.3 X-ray Mapping

X-ray mapping can quickly provide information about the distribution of different elements on an image of the sample.

These advantages and disadvantages are very general because there are too may variables in the types of specimen, the types of analysis, and the local concentrations of the elements. However, there is one big advantage of taking this overview, as it has an important influence on the way samples are prepared and analyzed. The importance of specimen preparation is the focus of the present book, whereas the operational procedures for x-ray microanalysis are extensively discussed in the book by Goldstein et al. (2004).

It has already been established that preparative methods designed to preserve structure alone create problems particularly with biological and wet specimens. For example, in a study on the distribution of aluminum in living tea leaves by Echlin (1996), the ambient-temperature wet chemical methods discussed in the previous chapter, caused a threefold to tenfold reduction in the amount of aluminum measured in different cell compartments compared with the amounts measured in samples stabilized using low temperature procedures. These low temperature methods are generally acknowledged to be the best way to preserve liquid, wet, and biological samples for elemental analysis. The best signals for x-ray microanalysis in the SEM are obtained from flat, highly polished, conductive samples with no preparation induced deformations.

9.3.5 Cathodoluminescence: Molecules and Macromolecules

Some materials, such as different insulators, polymers, semiconductors, and even biological samples when irradiated by high energy electrons will emit long wavelength photons (200–900 nm) in the UV, visible light, and IR regions of the electromagnetic spectrum. This phenomenon is known as cathodoluminescence (CL) and in some cases is bright enough to be observed by eye in an SEM equipped with an optical microscope. Figures 9.5 and 9.6 show a CL detector and an image of a specimen made using this method.

When an energetic beam scatters inelastically in some materials, electrons from the filled valence band are promoted to the conduction band, which creates an electron-hole pair. If there is no bias on the sample to separate the electron-hole pair, the electron and hole recombine and the excess energy is released as a photon of electromagnetic radiation. In many cases (direct gap materials) the

Figure 9.5. The Gatan MiniCl cathodoluminescence imaging system for use with a scanning electron microscope. The highly sensitive minature detector can be mounted inside the SEM at the end of a retractable probe. Picture courtesy Gatan, Inc. Pleasanton, California, USA

Figure 9.6. An SEM cathodoluminescence image of a polished piece of granite from Llano, Texas. The bluish cross hatched area is potassium felspar and the purple–red area is sodium felspar. Picture courtesy Juergen Schieber, Department of Geology, Indiana University, USA

electrons fall directly back across the band gap producing radiation peaks at specific well-defined energies that are characteristic of the material. In other cases (indirect gap materials) the recombination is mediated by impurity stages, and spectral characteristics (wavelength, line width, etc.) are strongly influenced by the presence of impurities.

Cathodoluminescence (CL) occurs naturally in certain solids such as semiconductors, minerals, rocks, and some biological samples. A book by Boggs and Krinsley (2006) gives details of how this technique may be applied to the study of sedimentary rocks. A paper by MacRae and Wilson (2008) gives an extensive luminescence data base of minerals and metals for ion luminescence and photoluminescence as well as cathodoluminescence. A recent paper by Smet et al. (2008) describes how an energy dispersive x-ray spectra may be used to spatially resolve cathodoluminescence in luminescent materials.

In contrast, plastics, glass, and some metals have a very weak CL signal. The poor emission of some specimens may be enhanced by the addition of phosphors and fluorescent dyes, which can emit up to 10% of the absorbed energy.

The CL technique provides a means of characterizing different compounds of material, but because the interactive volume that gives rise to the cathodoluminescence is rather large, the technique does not have a very high spatial resolution. The emission spectra are quite broad and overlap at ambient temperatures but at low temperature, the spectra are narrower and more clearly defined. Generally, the CL signal varies with the primary beam energy, and the cutoff energy typically depends on the sample surface type. High energy currents are preferred for high CL intensity, but this may damage sensitive specimens, although this can be ameliorated by working at low temperatures. The best signals for analysis in the SEM come from flat, conductive samples with a minimum of preparation-induced deformations.

These first five ways of obtaining chemical information about samples are generally non-invasive inasmuch as they usually do not need additional chemicals to be added to the sample to obtain the analytical information. As shown later, it may still be necessary to stabilize the specimen in order to obtain the information from the sample and ensure they are conductive. In contrast, the remaining five methods are invasive inasmuch as it is necessary to add chemical to the specimen in order to obtain meaningful analytical information.

9.4 Methods in Which Specimen Need Chemical Intervention Prior to Analysis

9.4.1 Quantum Dots: Elements and Molecules

These recently developed materials provide a means of labeling specific sites in specimen. They are manufactured from semiconductors such as cadmium selenide (CaSe), lead selenide (PbSe), gallium arsenide (GaAs), and indium arsenide (IrAs), which can be formed into very small particles ranging in size from 1 to 20 nm. The remarkable feature of these particles is that they emit different colored photons depending on their size. Small particles emit blue light and large particles emit red light with a wide range of different colors in between. For example, one material at 3 nm diameter quantum dots emits green light (520 nm); the same material of 5.5 nm diameter emits red light (620 nm). The quantum dots, which are inside a protective outer shell of zinc sulfide, are strongly hydrophobic, but can be made hydrophilic by coating them with a monolayer of a phospholipid micelle that enables the quantum dots to be suspended in water and, with the appropriate technology, be attached to antibodies in the same way that colloidal gold particles are attached to antibodies (see later).

Biologically active quantum dots have a number of distinct advantages over the more conventional fluorescent tagging techniques used in light microscopy. The emitted light has a very narrow emission band which is very bright, is photo-stabile (i.e., does not fade with time) and is non-cytotoxic.

Quantum dots are being used in the same way that fluorescent tags are used in localizing materials in biological samples; they can also be used in the SEM for non-biological material. David Joy (private communication, 2008) has been able to get a good cathodoluminescence signal from an aqueous suspension of quantum dots in an SEM wet cell at 20–30 kV and a beam current as

9.4 Methods in Which Specimen Need Chemical Intervention Prior to Analysis 197

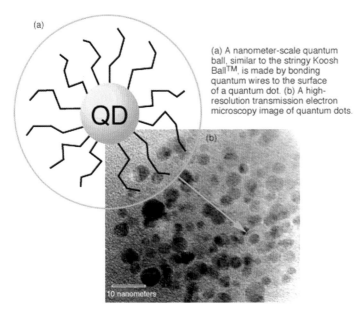

Figure 9.7. Diagrammatic representation of the form of a quantum dot. (a). A 3 nm diameter inorganic material quantum dot imaged in the electron microscope (b). Picture courtesy Howard Lee, Lawrence Livermore National Laboratory, California, USA

high as 100 pA. Figure 9.7 shows the way quantum dots are assembled and used in imaging.

9.4.2 Radioactive Labeling Methods: Molecules and Macromolecules

Autoradiography is an analytical procedure in which specific radioactive molecules are incorporated into known synthetic pathways to reveal their location in the sample or follow their synthesis, turnover, and movement. The presence of the radioactive label is revealed by the development of silver grains in a thin layer of photographic emulsion placed of the parts of the specimen into which the radioactive label had been incorporated.

The technique was popular for a number of years for studying both sections and flat surfaces of biological material using both light and transmission electron microscopy. Because of the rougher surfaces of many samples, it took a longer time for the method to be used with the SEM. The sample preparation is elaborate and time consuming and is generally not considered a high resolution analytical technique. For these and other reasons, autoradiography is now little used in the SEM and has, to a large extend, been superseded by other more sensitive analytical procedures. The paper by Chang and Alexander (1981) should be consulted for more information on autoradiography and the closely related topic of autofluorography.

9.4.3 Staining: Molecules and Macromolecules

Staining and histochemistry have been used for more than 200 years to identify and characterize the chemical composition of both living and non-living material. Stains are either positive or negative. The positive stains interact with the sample, whereas negative stains do not react with the sample but add contrast to the background surrounding the material.

The identification depended on changes in color initially seen with the naked eye and then with varying forms of light microscopy. Unlike light microscopy, which depends on differences in wavelength contrast to identify different regions in the sample, the SEM depends on differences in atomic number contrast as one way to obtain chemical differentiation. The development of both positive and negative "stains" for electron microscopy has developed from the earlier light microscope stains and the following general references contain information that can be applied to the development of techniques for use with the electron microscope generally and the SEM more specifically (Bancroft and Cook, 1984; Bancroft and Stevens, 1977; Hayat, 1975; Locquin and Langeron, 1978; Pearse, 1985).

- *Positive staining* is an important procedure to increase image contrast, and a number of different ways to do this were discussed in the previous chapter.
- *Negative staining* with heavy metal salts were used primarily in high resolution TEM but a very recent paper by Massover (2008) shows that light element salts of Na, Mg, and Al can be used as effective negative stains for structural biology.

The positive staining methods used for sample analysis involve introducing *identifiable* heavy metal ligands to specific chemical sites in an otherwise lighter element matrix in order to increase the electron scattering in these regions. The location of the stain signal may be indicated by BSE and confirmed by XRP. There are two different ways staining may be carried out. A recent and rather intriguing paper by Sata et al. (2008) gives details of how an Oolong tea extract may be used to stain sections!

1. A procedure that relies on the unique specificity of a chemical ligand *within* the sample to a solution of a specific heavy metal salt applied from the *outside* of the specimen. For example, silver ($Z = 47$) salts can be used to localize solutions such as sea water, which contain chlorine. Figure 9.8 is an example of this approach.
2. The second approach is less specific and reveals the presence of chemical groups in the specimen. For example, ruthenium tetroxide ($Z = 44$) can be used to locate aromatic rings and ether alcohols in some polymers; lead salts ($Z = 82$) can be used to locate phosphates in biological material and tungsten

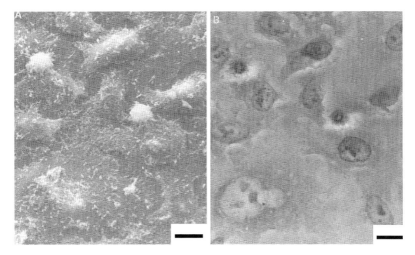

Figure 9.8. Secondary electron image (a) and a reverse contrast BSE image (b) of HeLa tissue culture cells stained with Wilder's silver stain. The silver selectively stains the chromatin and perinuclear regions of the cells. Magnification marker = 10 μm. Goldstein et al. 2004

($Z = 74$), and phosphotungstic acid can be used to localize hydroxyl and carboxyl groups. These procedures are generally only useful for a broad group of chemicals.

9.4.4 Histochemistry: Molecules and Macromolecules

This is more specific than staining and is designed to link the end product of an enzymatic reaction within a biological specimen to a heavy metal salt in an attempt to localize regions of specific metabolic activity. For example, diaminobenzidine (DAB)-osmium can be used to localize oxidases and Table 9.2 shows how cobalt salts can be used to alkaline phosphatase enzyme activity.

Both staining and histochemistry procedures are not without their problems, particularly with biological samples. The two methods are only satisfactory if the reactions are known to be quite specific with no cross-reactivity and the reaction products remain localized and unaffected by other parts of the total processes of sample preparation. At best, the two methods only give a general impression of the location of a particular ligand, and the spatial resolution is only marginally better than that obtained with a light microscope. Of course, care must be taken in interpreting the results.

9.4.5 Immunocytochemistry: Macromolecules

Although this technique is primarily for biological material, it does have a spinoff to other types of samples such as dry organic material such as wood and natural fibers. For readers unfamiliar with this technique, the book edited by Renshaw (2007) provides a useful introduction to this rapidly developing technology.

Table 9.2. An example of how enzyme histochemistry can be used in the SEM to localize regions of metabolic activity. The enzymes alkaline phosphatase with the substrate β-glycerophosphates and in the presence of calcium ions forms calcium phosphate. The calcium phosphate in the presence of cobalt ions and ammonium sulphide forms a brown, insoluble precipitate containing cobalt which can be localized either by BSE or x-ray photons

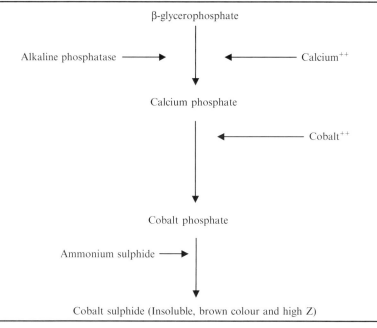

Antibodies, lectins, and immunoglobulins have a high affinity and specificity for many organic chemical groupings in biological macromolecules and smaller molecules. Antibodies are proteins and can only be seen directly in their native state at very high magnifications. They are, however, readily located in cells and tissues at lower magnifications when attached to a suitable label such as a fluorescent dye, which can be seen in the light microscope or a heavy metal such as gold particles, which may be imaged in an electron beam instrument. Immunocytochemistry using colloidal gold particles as markers is the best in situ analytical method for use with biological specimens. The best signals for analysis in the SEM come from BSE images from reasonably flat, conductive samples with recognizable preparation induced deformations.

9.5 The General Rules Dictating Satisfactory Sample Stabilization

Structure and composition are closely conjoined, and the principles underlying any preparative techniques for preserving the chemical composition of a sample must conform to the following guidelines:

1. The preparative methods must be able to retain elements, molecules, and macromolecules at their original concentration and location in the sample.
2. The methods should not remove from, or add to, the natural chemical composition and identity of the sample.
3. Any proposed preparative techniques, although biased toward retaining chemical information, also should retain recognizable structural components.
4. The best methods are usually the least invasive.
5. The best-prepared sample is the one that remains closest to its natural state.

Most of the early stabilization techniques were derived from the wet chemical procedures developed for specimens to be imaged by light and transmission electron microscopy. This created problems, for as the previous chapter showed, the emphasis was primarily on preserving macromolecular structural features, generally at the expense of differential solubilization, relocation and loss of elements in the specimen. It was also shown that some types of specimens, i.e., metals and hard dry inorganic materials, needed little or no sample stabilization, in contrast to biological and wet samples that needed a very different approach to the stabilization procedures in order to preserve their chemical identity and their structure.

9.5.1 The General Strategy for Sample Preparation

Before discussing these different strategies, it is important to appreciate that samples can be analyzed qualitatively or quantitatively.

- *Qualitative* analysis is used to determine the *presence* and general spatial distribution of a substance in a sample, i.e.: Is potassium present in one part of a rock but not in another? In many cases it is just a simple yes or no question, and may not require an elaborate sample stabilization procedure.
- *Quantitative* analysis is used to measure the *amount* of a substance and its specific spatial distribution in the sample, i.e.: What are the distribution, location, and local concentration of chromium, cobalt, and manganese in a newly acquired left knee prosthesis? This type of analysis requires careful stabilization procedures.

Any stabilization protocol must combine the high spatial acuity of secondary electron images with the procedures used for chemical detection. Bearing in my mind the wide variety of samples that are of interest, it is first advisable to make a list of the parameters that might affect the analysis before preparing the sample. The following questions should be answered before designing the experimental protocol for attempting to stabilize a particular specimen for analysis:

- What type of analytical instruments and methods of imaging are available?
- What information is being sought in the sample?
- What information is already available about the structure of the sample?
- What information is already available about the location of the chemical(s) of interest?
- What information is available about the binding state of the constituent elements?
- What are the normal hydration properties of the sample?
- What structural and analytical resolution is required?
- What degree of analytical sophistication is required?

9.5.2 Criteria for Judging Satisfactory Sample Stabilization

The experimenter must set up various criteria for judging whether or not a preparative procedure has worked. Satisfactory structural information usually is not enough and the following polemic should be considered:

1. The stabilized structural details should be preserved at a spatial resolution ten times better than the expected location of the region to be analyzed.
2. It must be possible to identify both the spatial location and the amount of chemical losses from the sample and translocation within the sample. This is usually very difficult to measure.
3. Avoid using any chemicals that will either mask or contribute to the signals from the chemicals being analyzed. This edict applies not only to the sample stabilization procedures, but also to all the other procedures and equipment used during the whole process of analysis. For example, do not use a phosphate buffer in a recipe for the analysis of bone, avoid using a gold:palladium conductive coating layer on polished garnet and do not use a copper support grid for the analysis of marine animals. A quick check on the K, L, and M x-ray energies of the elements will reveal where potential problems may lie.
4. Measure any changes that occurred in the inorganic and organic solid matrix of the sample and the suspended and soluble material in the aqueous phase during sample stabilization. These changes may affect the accuracy of the analytical results.
5. Assess any and all information available from correlative studies that would help evaluate the effectiveness of the preparative procedures.

It is suggested that a pencil and paper are the first tools to use when designing a satisfactory stabilization protocol for chemically analyzing a specimen in the SEM. The reasons for carrying out the analysis should be unambiguous. A list should be made of all the chemicals considered for use during the different phases of sample preparation, and a careful check made that this list does not conflict with chemicals that might be in the sample.

9.5.3 Metals, Alloys, and Metallized Specimens

Principal approaches used for analysis are x-ray microanalysis, backscattered imaging and electron backscattered diffraction.

These types of specimens need little or no stabilization prior to analysis, but the best analysis is carried out on clean, flat, and highly polished surfaces. Although the general procedures for polishing and cleaning are discussed in Chapters 6 and 10, it is appropriated to briefly re-consider these matters.

9.5.3.1 Mechanical Polishing

The samples and their accompanying standards should be flat-polished, with the final polishing routine only leaving small scratches no deeper than 100 nm. There should be no visible scratches when the thoroughly cleaned sample is viewed in an optical microscope. The necessary highly polished surface can be easily achieved with pure metal and alloy sample using one or more of the standard metallographic polishing techniques described in Chapter 6. The standard metallographic techniques can also be used for preparing samples for EBSD. J. Michael (personal communication, 2008) recommends using vibratory polishing with a very fine slurry as the final step. PDJ Vibro Ltd. (www.vibratoryfinishing.co.uk) manufactures equipment for vibro-polishing. Great care is needed when using the electrolytic grinding and electro-polishing mentioned in Chapter 6 to ensure there is no selective removal or addition of chemical to the sample surface.

9.5.3.2 Ion Beam Milling

Ion beams can be used in two general ways to erode and clean the surface of hard resistant materials. The narrow focused ion beam (5 nm) used in an ion-beam microscope can also be used to mill specimens, clean surfaces, and reveal underlying layers for subsequent analysis. The advantage of precision ion milling is that very small areas of the sample can be localized and cleaned.

The ion beam thinning devices use a broad non-focused beam of ions to thin materials by sputtering away the surface of the sample and slowly remove layers of the target by bombarding it with ionized gas molecules. Typical thinning rates are on the order of several micrometers per hour.

The difficulty with these techniques is to accurately judge the rate at which the specimen is removed to produce a highly polished, featureless surface. However, the ion beam methods are carried out in a vacuum and generally do not use chemicals. See Chapter 6 for more details.

9.5.3.3 Handling Surface Oxides and Corrosion

Difficulties begin to arise with metallic specimens that have surface layers of oxide and corrosion. If these specimens are to be chemically analyzed, they must not be etched or chemically treated after the very flat shiny surface has been achieved. The etching and chemical treatment can alter both the chemical composition of the metal surface and create height difference between different phases of the metal. Care has to be taken with passivation techniques that use chemical agents to make a polished metal surface inactive or less active to corrosion, or use de-passivation techniques that remove existing oxide layers and allow new oxide layers to form. Unlike the polishing and ion beam milling methods mentioned in the preceding, the chemical etching designed to reveal particular topographic substructures such as crystalline structures and grain boundaries, may cause changes in chemical composition.

9.5.3.4 The Final Cleaning

All traces of the abrasives used for sample polishing and any other chemicals used to prepare the sample surface should be removed by scrupulous cleaning (see Chapter 10). As an added precaution, a list should be made of the elements present in any chemical used for sample preparation, so that any unexpected x-ray peak artifacts may be identified.

9.5.4 Hard, Dry Inorganic Samples

Principal approaches for analysis in the SEM are x-ray microanalysis, backscattered imaging, electron backscattered diffraction, and quantum dots. The preparation of these materials is similar in many respects to the methods used on metals. These samples need little or no stabilization.

9.5.4.1 Ceramics and Geological Samples

Care must be taken in the selection of the polishing compounds as many mineral samples are quite hard. Diamond pastes are commonly used for the final polishing. The samples may require etching with solutions contain HF to allow microstructural features to be imaged and analyzed. Thermal etching can be used to produce surfaces suitable for microstructural imaging. Electron backscattered diffraction (EBSD) is commonly used to discriminate among phases in the specimens, and the appropriate specimen preparation techniques should be used.

9.5.4.2 Microelectronic Devices, Integrated Circuits, and Package Devices

These types of samples are made from metals, ceramics, and polymers and consequently need to combine different preparation methods, as demonstrated by Bousfield (1992). The first step is to expose the integrated circuit and care has to be taken when opening the device to reveal the areas or components of interest. Once the cover has been removed, it is prudent to fill the exposed interior with a liquid epoxy resin that is carefully allowed to polymerize and harden under vacuum. This keeps the different components in place.

For example, if a detailed analysis is required of the various metal layer and passivation thicknesses, the resin embedded material may be sectioned with a diamond saw and the exposed interior mechanically polished. The components are difficult to polish because of the wide range of hardness, and the processes combine the characteristics of the techniques used for metals, geological samples, and ceramics. It is usually prudent to use careful polishing with light loads and soft cloths. Chemical etching and ion beam etching may be used to reveal additional details. Modern microelectronic devices are very small and there is a move to avoid any wet chemical etching and cleaning during sample preparation and to use the high energy beam etching technique such as plasma and ion beam etching methods described in Chapter 6.

9.5.4.3 Semiconductors

These devices are composed of many different layers of many different insulating, conductive, and semi-conductive material such as polycrystalline light element compounds, oxides, and metals. Although the different components, to some extent, can be removed by differential chemical etching, these methods have been replaced by the plasma and ion beam etching processes discussed in Chapter 6.

9.5.5 Hard or Firm, Dry Natural Organic Material

Principal approaches for analysis in the SEM are x-ray microanalysis, staining, histochemistry, and to a limited extent, immunocytochemistry. These samples need little or no chemical stabilization prior to analysis except to carefully re-hydrate samples that are to be analyzed using immunocytochemical methods. The specimens in this group of materials are composed primarily of complex macromolecules of light element and, in order to analyze them in their natural state, they need little sample preparation other than charge elimination. One characteristic feature of these materials is their very low water content. This natural anhydrous condition is both an advantage and a disadvantage when it comes to specimen analysis. The advantage is that the specimens are reasonably stable and the only preparation that is required is to ensure that the surface is as smooth and flat as possible. If necessary, careful drying should be carried out using some of the methods described in Chapter 7. The disadvantages of these naturally dry specimens is that it is very difficult to effectively use any analytical procedures that rely on the addition of aquatic stains and histochemical.

9.5.5.1 X-ray Microanalysis

9.5.5.1.1 Naturally Occurring Light Elements
Depending on their local concentration, it is possible to obtain quantitative data about the elements that make up the samples. The abundance of the elements H, C, O, and N suggests that the best approach to analyze is first to use the well-tried standard chemical methods before using x-ray microanalysis. The other minor light elements (Na, Mg, Al, Si, P, S, Cl, K, and Ca) usually may be detected and, if in sufficiently high local concentration, be quantitatively analyzed using x-ray microanalysis.

The natural concentration of the common trace elements (Cr, Mn, Fe, Co, Cu, and Zn) is usually so low that they cannot be localized and identified using x-ray photons. However, sometimes it is possible to detect and quantify unusually high local elemental concentrations in some unusual plants and animals. Lead in some dried grasses that grow on the Lizard Peninsula in Cornwall, aluminum in dried tea, copper in the roots of the copper flower *(Becium homblei)* found in the Zambian copper belt, and the quite high levels of silver in the silver flower *(Erioganum)* that grows in Montana. Sulfur can be detected in some hairs and wools.

9.5.5.1.2 Artificially Added Elements
Wood preservatives contain a variety of both organic chemicals and some heavy metals such as arsenic, copper, chromium and cadmium. Figure 9.9 below shows a wood sample treated with a preservative. Their location and local concentration can be detected in treated timber and their products. In the same way titanium can be localized in high quality white paper.

9.5.5.2 Staining and Histochemistry

The various chemicals used to localize different active ligands in dry organic material are only effective in a hydrated state. The following example shows the limits of using staining chemical on dry specimens. Dry seeds contain a number of different carbohydrates, including starch, cellulose, pectins, and lignin and may be stained with chemicals that contain high atomic number elements. The light microscope, for example, will show that dilute solutions of iodine (53) stains starch a

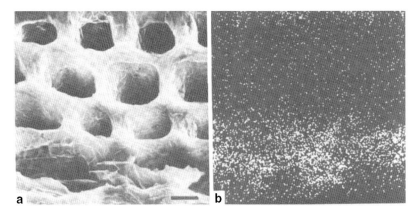

Figure 9.9. A piece of pine wood impregnated for three days with a copper-containing preservative. a) SE image showing the vascular tissue across the bottom of the image. Bar marker = 40μm. b) X-ray map for Cu Kα of the same area showing that most of the copper preservative is in the vascular tissue

light blue; ruthenium (44) stains pectin red-brown; zinc (30) stains cellulose violet; and molybdenum (42) stains pectins an intense blue. However, the heavier elements in these stains are at such a low concentration that although they are effective stains for light microscopy, this approach is generally ineffective for image and chemical differentiation in the SEM.

The use of stains and other related materials in connection with sample analysis is a mixed blessing. It is necessary to balance the artifacts that may arise from the use of invasive wet materials on dry samples, with limited spatial resolution and usefulness of only finding out that leather contains protein and that plant waxes contain lipid polymers. In summary, staining and histochemistry are not high resolution analytical techniques for dry organic materials.

9.5.5.3 Immunocytochemistry

Although these methods are vitally dependent on using wet chemical processes, this disadvantage is offset by the high spatial resolution when used to localize complex chemical groups in dry organic specimens. The following example shows how useful this method is in plant biology. Arabinoxylanes, which are one of the hemicelluloses, are the major non-starch polysaccharides in the dry outer cell walls of cereal grains such as wheat. The arabinoxylanes influence the rheological function in dough formation, and hence the quality of bread making. One example of a very specific method of stabilization is the study carried out by Jing Xia, a Ph.D. student who worked in my laboratory and with the Cereals Innovative Center, DuPont Ltd. in Cambridge. The details of this procedure are discussed later in this chapter.

9.5.6 Synthetic Organic Polymer Materials

Principal approaches for analysis in the SEM are backscattered imaging, x-ray microanalysis, and chemical staining. These samples need little or no chemical stabilization prior to analysis.

These materials share some of the characteristic of the previous group of specimens. They are essentially composed of carbon and hydrogen with a few other light elements and, occasionally, high atomic number inclusions. The book by Sawyer et al. (2008) contains a wealth of useful methods for preparing samples for subsequent analysis in the SEM.

The other approach to the analysis of polymers is to identify, localize, and rarely quantify "natural" heavy metal inclusions within the polymer matrix. This approach cannot be classed as a non-invasive procedure because most of the plastics and polymers have been synthesized from natural gas and/or crude oil.

9.5.6.1 Backscattered Imaging

The backscattered electron, sometimes compositional images can be used to reveal the presence of high atomic number inclusions in flat samples of polymers. This approach is particularly useful when studying metal loaded fibers. Figure 9.10 compares two different images of a mineral-filled polymer. The upper SEM image provides no information about the chemical nature of the filling material, but the lower BES image shows the filler material as a bright band of higher atomic number material at the fiber edge.

9.5.6.2 X-ray Microanalysis

Nearly all polymers and plastics are very beam sensitive, which makes it difficult to carry out analysis in the SEM, because the high beam currents can cause rapid mass loss during analysis. Local concentration of a heavy metal in a polymer used to make a car tire are easier to analyze than where the same heavy metal is evenly dispersed throughout the polymer. Some qualitative analysis can be achieved provided the x-ray count rate of the element being analyzed does not decrease during analysis, as this indicates progressive mass loss. Quantitative analysis is even more difficult because of the problems of differential mass loss of both the element(s) being analyzed and the organic matrix in which they are embedded. An alternative to point x-ray analysis is to use some of the more recent software developed for x-ray dot mapping methods.

9.5.6.3 Staining

Staining of polymers is reviewed in detail in the book by Sawyer et al. (2008) and the previous chapter of this book, which gives an outline of some of the different stains that can be used to enhance

Figure 9.10. Secondary electron (a) and backscattered electron (b) images of a mineral–filled polymer composite. The BSE image shows only that the mineral filler is of higher atomic number than the polymer (Picture courtesy Sawyer et al., 2008)

the contrast of polymer images examined in the SEM. These same stains, to a limited extent, can be used to localize and characterize regions of chemical difference within a given polymer. Polymers and plastics are slightly more chemically active than the hard dry organic material considered in the previous section, and Table 9.3 shows how different functional groups of polymers may be selectively stained using heavy metal solutions. For example, Figure 8.1 in the previous chapter shows how chlorosulfonic acid can be used to reveal the electron-dense inter-lamella surfaces of polyethylene.

The difficulty with using stains for polymers is that, although they are excellent for increasing image contrast, they are much less specific than the stains used for analyzing biological material. There are very few heavy metal stains that will stain only one type of polymer functional group. For example, tin chloride has an affinity for the -CONH- groups in nylon.

Table 9.3. Heavy metal stains used to characterize different polymer functional groups

+ Polymer	Stain
Unsaturated hydrocarbons, alcohols, ethers and amines	Osmium tetroxide
Unsaturated rubber	Ebonite
Polyethylene and polypropaline	Chlorosulphonic acid plus uranly acetate
Amines and esters	Phosphotungstic acid
Aromatics, bi-phenols and styrene	Ruthenium tetroxide
Esters and aromatic polyamines	Silver sulphide
Acids and esters	Uranyl acetate

After Sawyer et al., 2008.

Sawyer et al. (2008) show that most of the stains used with polymers are positive stains that result in a specific region of the polymer showing increased electron emission. In negative staining, the shape of small particle on a flat surface is revealed by staining the region *surrounding* the particles. The following four examples, taken from Sawyer et al. (2008), show how heavy metal stains can provide some chemical information about polymers.

1. *Osmium tetroxide*. This reactive and toxic chemical, which can be used either as a vapor or in an aqueous solution, reacts with the C=C bonds, which are a general feature of many polymers. There is contrast enhancement in unsaturated latex particle when treated with osmium tetroxide and it also causes preferential absorption in the amorphous regions of spherulites in semi-crystalline polymers. Pre-treatment with boiling alkaline saponification solutions may help to enhance the staining ability of osmium tetroxide.
2. *Ruthenium tetroxide*. This is a powerful oxidizing agent that oxidizes aromatic rings and can be used to stain saturated polymers and rubbers. It also may be used to stain latex and resins by cross-linking ester groups. One percent aqueous solutions of ruthenium tetroxide have an affinity for ethers, alcohols, aromatic, and amino groups. This stain does not have a high specificity.
3. *Phosphotungstic acid*. Solutions of this chemical can be used to reveal the lamella structure of nylon fibers and the lamellar regions of polypropylene. The chemical also reacts with surface functional groups such as hydroxyl, carboxyl, epoxide, and amine.
4. *Uranyl acetate*. Aqueous solutions can be used to post-stain polyethylene after prolonged pre-treatment with chlorosulfonic acid. The same chlorosulfonic acid-uranyl acetate treatment can be used to stain the amorphous regions of semi-crystalline polymers such as poly-olefins.

The *term staining in polymer analysis* is used rather widely in the same way that the term is used in the previous group of hard, dry organic materials. The staining provides information at the macromolecular level and not, as shown in the next section, at the molecular level.

The four groups of specimens discussed so far generally do not need extensive sample stabilization prior to chemical analysis. Like all specimens, they need constant cleaning, and an assurance that what little treatment has been made before analysis has not altered the natural content and distribution of their chemical identity.

The final two groups of specimens are, as they say, a different kettle of fish! The previous chapter showed that care has to be taken stabilizing biological, wet, and liquid samples in order to retain their structural integrity. These practices are even more demanding when it comes to retaining their chemical integrity.

9.5.7 Biological Organisms and Derived Materials

Principal instrumental approaches to analysis are backscattered electrons and x-ray microanalysis. The use of backscattered electrons to analyze biological specimens will not be discussed here. For at best they can only indicate whether groups of high element materials are present in a low element material matrix.

Specimens in this group are hydrated, sensitive to radiation damage, structurally fragile, extremely susceptible to biochemical perturbation and, as a consequence, are difficult to prepare, properly, for in situ analysis of macromolecules, molecules, and small groups of atoms. In addition to this catalog of characteristic features, biological material is living and dynamic.

All living biological tissues, cells and solutions have diffusible movements and some cells have cytoplasmic streaming. These movements are quite fast. Electrolytes in solution move about 1 µm in 0.5 ms, although it is estimated the rate slows to 0.1 µm in 0.5 ms in cell cytoplasm. Physiological events occur within 1–5 ms and many chemical events are virtually instantaneous. In contrast, the diffusion rate of many of the chemicals used for sample stabilize are of the order of 1 µm/s and even cryo-stabilization at 1 µm/20 ms may be too slow to arrest some procedures. These relatively slow diffusion rates of the chemicals used in sample stabilization may be an advantage for preserving *structural features* but they can cause serious problems in the analysis of the physiochemical matrix of living tissues.

Samples of biological material can be divided into five different groups in respect to their interaction with electron beam instruments:

1. *Bulk samples*, which do not transmit the electron beam, are limited to an analysis of natural surfaces, polished surfaces, smooth(ish) fractured surfaces, and sectioned and planed

surfaces. The x-ray spatial resolution is between 50 nm and 10 µm.
2. *Isolated cells and large organelles,* although only 5–20 µm thick, are usually treated as bulk samples. The x-ray spatial resolution is between 50 and 500 nm.
3. *Particles and fibers* that may have been eluted, collected, or analyzed in situ; the x-ray spatial resolution is between 50 and 100 um.
4. *Thick sections* up to about 2-µm thick in which reasonable morphological details may be resolved in the scanning transmission mode, the x-ray spatial resolution is between 30 and 80 nm.
5. *Thin sections* up to 200 nm thick in which high resolution morphological detail may be resolved, the x-ray spatial resolution may be as good as 3–20 nm.

There are a number of different ways to stabilize the rather delicate biological samples so they retain their in situ chemical information in addition to their structural properties. The various methods that will be discussed are refinements of the procedures discussed in the previous chapter which must also fulfill the following criteria:

1. The stabilization methods must retain elements, molecules, and macromolecules at their original concentration and their original natural location in the sample.
2. Procedures and their associated chemicals should not remove or add to the original chemical identity of the material being analyzed.
3. The methods, although biased toward retaining chemical information, must also retain recognizable structural components.
4. The best methods are usually the least invasive.
5. The best method is the one that retains the sample closest to its original state.

9.5.8 Environmental SEM

Nothing is done to the specimen, and the liquid environment remains unchanged. This approach, although very important, falls outside the remit of this book and is not considered further. The fact that nothing is done to change the aqueous environment of the biological samples means that the movement of dissolved materials and diffusion processes are un-incumbered and would make it difficult to measure the precise location of a chemical entity.

There are two main approaches to stabilize biological specimens in order to try to retain and measure their natural in situ location of chemicals:

1. *Ambient-temperature wet chemical methods.* The natural chemical and liquid environment of the sample is chemi-

cally stabilized to varying degrees. This approach compromises all five criteria discussed in the preceding and is generally no longer considered a satisfactory approach. The shortcomings and consequent hasty dismissal of the approach will considered only briefly.
2. *Low-temperature SEM.* Only the phase of the liquid environment is changed. This approach meets all five of the criteria given in the preceding. The rationale of the low temperature approach has been discussed in previous chapters and is only considered in his chapter where the methods have a particular relevance to chemical analysis.

9.6 Ambient-Temperature Wet Chemical Stabilization Methods

All chemical stabilization procedures cause changes in sample permeability, and our concern must be to seek ways to staunch the flow of the soluble and suspended fine particles from and within the specimen. A careful study of the literature over the past 40 years shows that there is a gross loss *and* redistribution of nearly all dissolved elements during all phases of conventional preparative methods. The losses are not uniform; for example, in one part of a sample, large amounts of material will be lost and in another part of the same sample these losses are not apparent. Table 9.4 gives some sense of the large losses of dissolved and suspended elements from specimens during these wet-chemical procedures.

Care must also be taken over the choice of buffers, some of which may contain the very elements being analyzed and others may contain elements that overlaps the x-ray signal of the element being analyzed. Table 9.5 gives some examples of these overlaps.

9.6.1 Mild Stabilization

Many attempts have been made to use so-called "gentle fixation" by using organic fixatives at low concentrations and low temperatures for short periods of time.

Dilute solutions of glutaraldehyde or paraformaldehyde in organic buffers such as piperazine-N-N'-bis(2-ethanesulphonic acid) (PIPES); N-(-hydroxyethylpiperazine-N'-2-ethane sulfonic acid) (HEPES); and tertiary amine-heterocyclic buffers, may lessen (but do not prevent) loss of material from the sample. Mild stabilization at 275–280 K using aldehydes and diamino-esters that cross-link proteins, may be useful for fairly well bound elements. Sadly, ambient temperature wet-chemical stabilization is too slow for use with diffusible elements and insufficiently specific for the subsequent analysis of most biological specimens.

The attempts to use solutions of heavy metal salts to either *precipitate* or *stain* dissolved elements in situ is now not generally

Table 9.4. Examples of elemental losses from plant and animal material after stabilization with different chemical solutions present in isotonic buffers. Flame photometry or liquid scintillation counters were used to analyze the material before and after stabilization. Information from the scientific literature over the past 30 years

Tissue	Element	Stabilization Chemical	Percentage loss/gain
Uterus	K	1% OsO^4	−75%
Red Blood Cells	K	2% formaldehyde	−70%
Smooth muscle	Na	5% glutaraldehyde	+200%
Aorta	Ca	3% glutaraldehyde	−50%
Pancreas	Ca	0.5% glutaraldehyde	−42%
Smooth muscle	Ca	5% glutaraldehyde	+35%
Tumor Cells	Mg	12% glutaraldehyde + 1% OsO^4	−55%
Bivalve tissues	Cd	1% glutaraldehyde	−46%
Broad bean leaf	Mg	2.5% glutaraldehyde	−60%
Leaf Tissue	Rb	1% OsO^4	−90%
Diatoms	Si	2% glutaraldehyde + 1% OsO^4	−35%
Leaf tissue	^{36}Ca	1% OsO^4	−80%

considered a useful method, for the following reasons, particularly for diffusible elements and those involved in active physiological processes.

9.6.2 Precipitation: Elements and Small Molecules

Twenty years ago a number of workers (Ellisman et al., 1980; Van Steveninck and Van Steveninck, 1991) attempted to precipitate diffusible elements in situ prior to chemical stabilization. The techniques are based on the reaction of an ion with a heavy metal compound to produce an electron-dense insoluble precipitate that can be imaged by BSE and/or characterized by x-ray microanalysis. For example, the chloride ion can be precipitated by silver nitrate. An alternative approach, in which the ion itself is a heavy metal, is to use an organic reagent as the precipitating agent. For example, ATPase can be used to precipitate lead deposits in biological samples. Table 9.4 shows some of the precipitation reactions that were applied to biological specimens.

The combined chemical stabilization and precipitation process can cause additional losses of material from the specimen, as Tables 9.5, 9.6, and 9.7 demonstrates.

Precipitation of diffusible and soluble elements is now no longer used for the following four reasons:

1. Some of the chemicals used as precipitation agents are toxic, and many of the protocols cause massive loss of material from the samples.

2. Some of the more popular precipitating agents, such as pyroantimonate, are not very specific and will precipitate sodium, magnesium, potassium, and calcium.
3. There is now good evidence to show that the precipitating agent acts as a focus for electrolytes and other dissolved materials and causes the diffusible elements to move from

Table 9.5. Potential overlaps between elements in the specimen and elements in some buffers commonly used for specimen stabilization

Buffer	Potential overlap with sample
Sodium cacodylate buffer	Arsenic La 1.28 keV overlaps magnesium Ka 1.25 KeV. The buffer also contains Na, CL and Ca which will interfere with natural elements
PIPES buffer	Contains sodium, phosphorus and chlorine
Various phosphate buffers	Contain sodium, phosphorus and potassium

Table 9.6. Precipitation reactions for *in situ* demonstration of soluble inorganic ions and the enzyme ATPase in plant and animal material

§§ Specimen	Ion or ATPase	Precipitating agent
Animal and plant	Cl^-	Silver acetate or lactate
Animal and plant	Na^+	Potassium pyroantimonate
Animal and plant	$Na^+, Mg^{2+} \cdot Ca^{2+}$	Potassium pyroantimonate
Animal	Ca^{2+}	Potassium oxalate
Animal	PO_4^{-}	Lead acetate
Animal	ATPase	Lead nitrate
Plant	Na^+, Ca^{2+}	Benzamine
Plant	Ca^{2+}	Ammonium oxalate

Table 9.7. Loss of natural elements and applied radioactive tracers from biological specimens after exposure to different stabilization protocols and precipitating reactions. All the processing regimes contained the appropriate isotonic buffers at the correct pH

Tissue	Soluble element	Processing	Percentage loss
Animal Uterus	Na^+, Ca^{2+}, Mg^{2+}	3% glutaraldehyde + 2% K-pyroantimonate	40 to 90%
Animal Membranes	Ca^{2+}	6% acrolein + 1% sodium oxalate	12%
Plant roots	^{65}Zn	1% glutaraldehyde + 30% sodium sulphide	10%
Plant leaves	^{203}Hg	0.3% glutaraldehyde + 3% sodium sulphide	70–80%
Plant leaves	^{36}Cl	1% osmium tetroxide + 30mM silver acetate	70–90%

Data from Morgan (1980).

their natural location in the cells and tissues to an unnatural deposit. A closely related problem is the phenomenon of stain aggregation. Not very much is known about the extent of this problem, which can be of concern for high resolution electron microscopy. Hayat (2000) considers that stain aggregations of more than 1 nm would be of concern for high resolution TEM and gives a stain aggregation figure of 4–6 nm for a silver compound.

4. It is very difficult to ensure that there is no movement of the precipitates during subsequent preparative techniques. Whereas Läuchli et al. (1978) were able to show that after preloading tissue with ^{36}Cl only about 4% of the chloride was lost during silver precipitation techniques, whereas Yarom et al. (1975) showed that the calcium precipitated with pyroantimonate was completely lost during subsequent alcohol dehydration. Warley (1997) shows that elements are also lost during sectioning, i.e., 87% of Na^+ and 55% of Cl^+ and Ca^+ were lost from the specimens. Similarly, during staining with uranyl acetate, 45% Ca^+, and 62% Ca^+ were lost when samples were post stained with phosphotungstic acid.

5. Some of the heavy metal precipitating agents have x-ray spectra that overlap the light element x-ray spectra and create difficulties with analysis. For example, the $L\alpha$ lines of antimony at 3.605 and 4.132 overlap the $K\alpha$ lines of potassium and calcium and the M series of lead at 2.346 and 2.502 overlaps the $K\alpha$ line of sulfur.

9.6.3 Staining: Elements and Molecules

It was shown in the previous chapter that both selective and non-specific heavy metal staining was useful in increasing the image quality of specimens examined in the SEM. This type of staining should not be used in studies that center on the x-ray analysis of biological material because of the overlaps between the x-ray spectra of the stains and specimen. Tables 9.8 and 9.9 shows the wide range of overlaps that may occur between the characteristic x-rays of elements used for stains and the characteristic x-rays of the light elements that make up the bulk of biological material.

9.6.4 Histochemistry: Molecules

The histochemical approach links the end product of an enzymatic reaction to a heavy metal salt and localizes areas of specific metabolic activity. These procedures are beset by the same litany of problems associated with staining. Care is needed during sample preparation to retain the enzymatic activity, and low temperature preparation techniques are the best approach. An example of this procedure is shown earlier in Table 9.2. Other methods involve using diaminobenzydine-osmium to localize oxidases, insoluble lead salts to localize acid phosphatases and esters and copper

Table 9.8. A comparison of the characteristic x-ray energies of 98% of the elements in biological material with a range of some of the elements which might be considered staining agents

	Elemental composition	Characteristic x-ray energies
Biological material	C (Z=6) to Ca (Z=20)	K series 0.277KeV to 4.038KeV
Middle density stains	Ti (Z=22) to I (Z=53)	L series 0.452KeV to 3.936KeV
High density stains	La (Z=54) to U (Z=92)	M series 0.833KeV to 3.545Kev

Table 9.9. Potential overlaps i.e. (**X-Y**) of x-ray emission lines of elements associated with the sample preparation of specimens containing light elements. Assuming an overvoltage of 2.5. Smooth bulk specimens, undergoing quantitative energy dispersive analysis to within +/ 3% of value measured by wet-chemical analysis

Voltage (Kv)	Element	Kα line (eV)	Lα line (eV)	Mα Line (eV)
1	C	262		
	N	392		
	Ca		349	
2	O	523 (**Cr-L**)		
	Cr		571(**O-K**)	
	Fe		704	
3	Cu		928	
	Zn		1004	
	Na	1041		
4	Mg	1254		
	Se		1379	
	Br		1408 (**Al-K**)	
	Al	1487 (**Br-L**)		
5	Si	1740		
6	P	2015 (**Pt-M,Au-M**)		
	Pt			2051(**P-K**)
	Au			2123(**P-K**)
	S	2308 (**Pb-M**)		
	Pb			2346(**S-K**)
7	Cl	2622		
8	Ag		2984	
9	K	3313		
10	Ca	4039		

ferricyanide to localize dehydrogenases. The books by Van Norrrden and Fredriks (1992) give details of how this approach has been applied to biological samples.

Histochemical staining relies on the specificity of a chemical ligand *within* the sample for a heavy metal salt solution applied from *outside* the sample. This approach can be used for the in situ localization of a specific chemical group, a specific metabolic or chemical process and for following chemical changes at a specific location. For example, ruthenium tetroxide can be used to localize aromatic rings and ether alcohols; lead salts for phosphatases and tungsten as phosphotungstate to localize hydroxyl, carboxyl, and NH groups. Figure 9.11 shows how a stain containing silver has been used to localize chromatin and perinuclear material in tissue culture cells. Histochemical staining is best done on thick specimens before and after a particular chemical treatment. Chapter 6 in the book by Hayat (2000) contains a large number of recipes that may be easily adapted for use in the SEM.

9.6.5 Immunocytochemistry: Macromolecules

Macromolecular analysis uses immunocytochemistry during which specific antibodies conjugated to colloidal gold or quantum dots are used to localize active sites within samples followed by BSE and SE imaging and characterization by x-ray spectroscopy. Figure 9.12 shows the high quality resolution information that may be obtained using this analytical technique.

Immunocytochemistry is a significant analytical strategy for biological specimens and important in the analysis of macromolecules in the SEM. A brief description of the procedure are given in the following, but for more details of this rapidly developing technique, reference should be made to the publications by Hayat (2002), Polak and Van Noorden (2003), Dabbs (2006), and the recent book edited by Renshaw (2007).

9.6.5.1 Antibodies, Lectins, and Immunoglobulins

These macromolecules have a high affinity and specificity for certain organic chemical groups in biological material. Antibodies are proteins and can only be seen directly in their native state at very high magnifications. They are, however, readily located

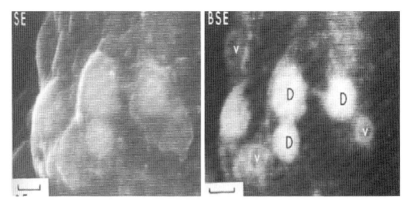

Figure 9.11. Secondary and backscattered images of rat bone marrow tissue showing the location of peroxidase activity stained with diaminobenzidine-osmium. Goldstein et al. (1992)

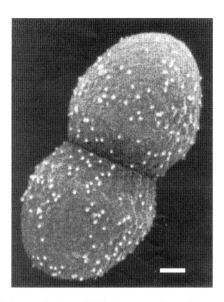

Figure 9.12. High-resolution BSE image of an immunogold-labelled (12nm) cell of the bacteria E. faecalies. Cell coated with 1nm platinum. Scale bar = 100nm. Picture courtesy S. Erlandsen et al. (2003)

in cells and tissues at lower magnifications when attached to a suitable label such as a fluorescent dye, which can be seen in the light microscope or a heavy metal such as gold, which may be imaged in an electron beam instrument. Immunocytochemistry, using colloidal gold particles as markers, is the best in situ analytical method for use with biological specimens. The best signals for analysis in the SEM come from BSE images from reasonably flat, conductive samples with a minimum of preparation induced deformations.

9.6.5.2 The Antibody–Antigen Reaction

The antibody–antigen interaction is well understood and characterized. From the immunocytochemistry point of view, a good antibody has a high affinity for its specific antigen.

Monoclonal antibodies are homogeneous and usually react with only one molecule and a single antigenic determinant (epitope). Monoclonal antibodies can be prepared in large amounts, but because of their high specificity, the epitopes may become cross linked during sample preparation that requires non-invasive stabilization procedures.

Polyclonal antibodies are heterogeneous and are usually directed against a number of different epitopes on the specimen that increases the chances of cross linking, but at the expense of specificity. The antibody–antigen interaction provide the remarkable specificity of the technique; the colloidal gold is the marker that provides the means of imaging the location of this interaction in the SEM.

9.6.5.3 Colloidal Gold Probes

The colloidal gold probes are gold spheres that come in a range of sizes from 2–40 nm and, when coated with a selected protein, can be attached to a specific antibody. The gold probes are electron-dense, have no known cytotoxicity, and are stable cytochemically. They can be used directly or indirectly to localize specific antigenic sites with a high sensitivity. Colloidal gold immunocytochemistry is very useful for examining natural or exposed surface antigens and for the three-dimensional examination of large areas of tissues.

9.6.6 General Features of Specimen Preparation

All the methods are based on using either a mild chemical stabilization and dehydration routine or some of low temperature techniques discussed in the next section. The preparation procedures fall into two general classes:

1. Pre-embedding techniques using the indirect procedure. If necessary, the inside of the sample is exposed by sectioning, ultra-planning, or smooth fracture and is then first incubated with the tissue-specific primary antibody followed by incubation with a secondary antibody conjugated to a gold probe. The specimens are then stabilized, dehydrated, and embedded in a resin. The procedure ensures good structural preservation and accurate immuno-localization of the epitope (the region of the antigen to which the antibody binds). The procedure is also good for external antigens on the surface of the specimen.
2. Post-embedding techniques using the indirect method. If necessary, the inside of the sample is exposed by sectioning, ultra-planning, or smooth fracture and the exposed surface is stabilized, dehydrated and, if necessary, embedded. The stabilized specimen is then incubated with the specific primary antibody followed by incubation with the secondary antibody conjugated to the gold probe. The procedure is suitable for both internal and external antigens.

A large number of different wet chemical stabilization and preparation methods have been developed for use with biological materials. Table 9.9 shows a generic wet chemical stabilization routine for immunocytochemical studies.

An example of a very specific method of stabilization and immunocytochemical localization is the study carried out by Xia (2001) a Ph.D. student who worked in my laboratory and with the Cereals Innovative Center, DuPont Ltd. in Cambridge.

The study involved the location of the carbohydrate, arabinoxylane, in the walls of the wheat caryopsis (grain) that play an important part in bread making. Antibodies were raised to a fragment of arabinoxylanes, which were then linked to a second antibody conjugated to colloidal gold particles. Figure 9.13 shows images of the cereal sample. Identifiable fragment, were imaged in the SEM using both SE and BSE signals. Her work showed that it was possible to localize the position and the varying amounts of arabinoxylane in the outer dry cell wall layers (pericarp) and the inner aleurone and endosperm layers of different wheat species and of other cereals.

The stabilization preparative technique which was used to prepare the images shown in Figure 9.13 are given Table 9.10.

9.7 Low Temperature Stabilization Methods

Low temperature methods are discussed in Chapter 7, where it is shown that they play an important non-invasive procedure in specimen dehydration and in Chapter 8, where they are one of

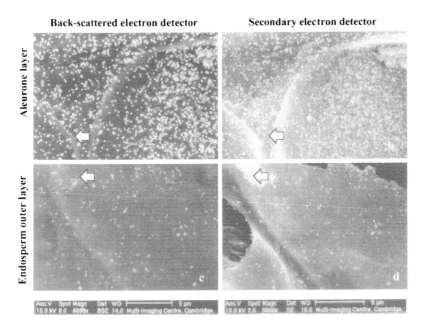

Figure 9.13. Immunolabeling of wheat arabinoxylem using colloidal gold-labelled antibody enhanced with silver. Two different regions of the wheat caryopsis were studied and a comparison made from, (a) and (c), using a solid state BSE detector and from (b) and (d) using an E-T secondary electron detector. The two images using the BSE detector have a marked atomic number contrast where as the two images using the SE detector have a diminished atomic number contrast but an increased topographic contrast. The four images show that there is considerably more arabinxylans associated with the cells of the aleurone layer than with the endosperm cell walls. Picture courtesy Jing Xia, University of Cambridge

Table 9.10. A generic preparation technique for the immunocytochemical analysis of biological samples. This basic method will need much additional experimentation to ensure there is a high specific signal and a low no-specific background

Stabilization	2% Formaldehyde + 0.05% glutaraldehyde in phosphate or PIPES buffer at an appropriate tonicity and pH for 12 hrs at 15°C. Applying microwaves during stabilization can help enhance expression of the antigens.
Dehydration	Progressive lowering of temperature dehydration to 80% ethanol.
Embedding	Slow infiltration in a hydrophilic acrylate resin.
Exposure	Polymerized resin is either sectioned, micro-planed or fractured.

the methods of stabilizing sample to obtain high fidelity images. Freeze drying and freeze substitution will be briefly reconsidered here as a means of stabilization specimen to retain their chemical content. A third method of low temperature stabilization is to do nothing to the sample except to carefully and rapidly quench cool the specimen and analyzing it in a frozen state at low temperature on the cold stage of the SEM.

9.7.1 Freeze Drying

The freeze dried specimens used for SEM images can also be used for chemical analysis provided they show no evidence of rehydration. The freeze dried samples for analytical investigations are best immediately embedded in a liquid plastic that is then polymerized. The drying and embedding must be a continuous process and the sample only exposed to the atmosphere once the embedding resin is fully polymerized. Details of different resin embedding procedures are given in Chapter 5.

An alternative procedure is to carry out controlled primary and secondary sample freeze drying in one of the separate cold stages, which may be bolted to the side of an SEM, which is fitted with its own cold stage. Quorum Technologies Ltd (www.quorumtech.com) produce the CryoSEM System PP-2000T and Emitech Products Inc. (www.empdirect.com) produce the Emitech K-1250 Cryogen Preparation System. The stages have two air locks and operate at a high clean vacuum. They are fitted with a heater and thermocouple for precise temperature control and enable pre-quenched cooled systems to be cryofractured, frozen dried, and coated with a suitable conducting layer.

The specimen is first quench cooled outside the microscope system and transferred under liquid nitrogen via an air lock, to the pre-cooled specimen holder of the accessory cold stage where it can be freeze dried. This procedure, which is only really feasible with small (1–2 mm^3) specimens, allows the frozen dried specimen to be metal coated before is transferred via a second air lock to the microscope stage.

This particular approach is the only sure way to ensure that the frozen specimen remains well below the ice re-crystallization temperature at all times and avoids the use of resin embedding processes that may introduce additional elements such as chlorine into the sample. Impurities may be introduced into the specimen if cryoprotectants are used in an attempt to reduce ice crystal damage during the initial quench cooling. Finally, it is necessary to make a judicious choice of materials to reduce charge elimination. These are matters we will return to in Chapter 11.

9.7.2 Freeze Substitution

This is a more complex process that has already been discussed in the previous chapter. The method has the potential of introducing impurities into different stages of the preparative procedure. The essential feature of freeze substitution is that water in the sample is dissolved at low temperatures in organic liquid that avoids the problems associated with ice crystal damage. Unlike freeze drying, the specimens invariably need some form of stabilization either before freeze substitution or during the process. A review of the current literature shows that freeze substitution is a very effective method of preparing all types of plant, animal, and microbial specimens, particularly in association with the immunocytochemical localization of macromolecules and large molecules.

The review by Skepper (2000) is a good introduction to some of the stabilization procedures associated with freeze substitution. Glutaraldehyde and formaldehyde are the most common stabilizing agents and Table 9.11, which is taken from this paper, shows a typical stabilization schedule.

Although freeze-substitution followed by immunocytochemistry is now one of the main ways of preparing biological

Table 9.11. Stabilization procedure for the immunocytochemical localization of arabinoxylans in the cell walls of the wheat caryopsis

1. The wheat caryopsis was cut into 1.5 mm thick sections.
2. The sections were first placed in a 0.1M phosphate buffer pH 7.4 containing 0.8% sodium chloride and 0.05% Tween 20 (PBST) for 2 hours.
3. The stabilization solution contained 0.05% glutaraldehyde in a 0.05 M PIPES buffer at pH 7.4.
4. The samples were briefly washed in the stabilization liquid and then soaked in the liquid for 1 hour at room temperature.
5. The stabilized samples were washed five times in the PBST solution given above.
6. The specimens were ready for the immunolabeling procedure.

From Xia, 2001.

Table 9.12. A potential stabilization solutions for use in association with freeze substitution for immunocytochemical analysis of biological samples

Chemical	Buffer	Time
1% glutaraldehyde(G)	0.1 m PIPES +2mmol L^{-1} $CaCl^2$	30–60 min
8% formaldehyde	0.1 m PIPES +2mmol L^{-1} $CaCl^2$	60–120 min
6% formaldehyde	0.1 m PIPES +2mmol L^{-1} $CaCl^2$	60–120 min
4% formaldehyde+ 0.1%G	0.1 m PIPES +2mmol L^{-1} $CaCl^2$	60–120 min
2% formaldehyde+0.05%G	0.1 m PIPES +2mmol L^{-1} $CaCl^2$	60–120 min
3% formaldehyde	0.1 m PIPES +2mmol L^{-1} $CaCl^2$	60–120 min
4% EDC	0.1 m PIPES +2mmol L^{-1} $CaCl^2$	60–120 min
PLP		60–120 min
1% formaldehyde	0.2 m PIPES + 3mmol L^{-1} $CaCl^2$	60–120 min

EDC is a water soluble carbodiimide.
PLP is periodate-lysine-paraformaldyde solution.
From Skepper, 2000.

material for chemical analysis, freeze substitution is occasionally used in association with the stabilization and preparation of light element compounds for subsequent x-ray microanalysis. Some typical stabilization schedules are given in Table 9.12.

It is not suggested that wet chemical stabilization procedures are routinely used on specimens containing diffusible and dissolved elements for freeze substitution for subsequent x-ray microanalysis. Unless the samples are very small, there should be concerned about the short times and low temperatures used with these procedures and whether the chemicals adequately stabilize the sample sufficiently well to avoid elemental losses. Element such as N and P, which are covalently bound in a macromolecular structure such as an enzyme, will suffer little elemental loss after wet chemical stabilization followed by freeze substitution to prepare samples for x-ray microanalysis. Data collected by Morgan et al. (1978) shows that there are losses of diffusible and soluble elements such as sodium but covalently bound elements such as nitrogen and phosphorous are retained. A summary of some of the evidence is shown in Table 9.13.

9.7.3 Frozen Hydrated Specimens

The third of the three low temperature stabilization procedures for chemical analysis is to do nothing other than initiate a phase change in the liquids, primarily water, in the sample. The physiochemistry of this process is discussed in Chapters 7 and 8, and in more detail in my book on low temperature microscopy and

Table 9.13. Some stabilization procedures used in association with freeze substitution for the analysis of biological materials using x-ray microanalysis

Tissue	Chemical	Temperature	Time
Plant material	0.3 to 1.0 glutaraldehyde in 0.1M phosphate buffer pH 7.4	278–283K	30–60 min
Plant material	2% formaldehyde in 0.1 M phosphate buffer pH 7.2.	277K	30 min
Microorganisms	2% glutaraldehyde in 0.1 M phosphate buffer pH 7.2.	288K	60 min
Animal material	5% glutaraldehyde in physiological saline pH 7.2	277–313K	2 hours
Animal material	2% acroline in veronal-acetate buffer pH 7.4.	280–300K	1 hour

analysis (Echlin 1992). The essence of the procedure is to quench cool specimens as fast as possible and then keep the sample below 123 K during subsequent processing and analysis. This approach depends on using low temperature preparation devices attached to an SEM fitted with its own cold stage. The chemical content of frozen hydrated specimens can only be identified and measured by x-ray microanalysis. It is most important to be certain that a sample is fully hydrated during analysis. Figure 9.14 and Table 9.14 shows the significant difference in the appearance and elemental concentrations between the frozen hydrated and frozen dried state.

At first glance, the best results should be obtained using the high pressure freezing equipment described in the previous chapter. This approach will produce frozen specimens, devoid of visible ice crystal artifacts, in thin (200 µm) samples examined in the TEM. It is more difficult to achieve this type of preservation in the SEM, which usually examine bigger samples and that, alas, may only be cooled slowly. Another problem is that thin frozen hydrated specimens while providing high quality images do not usually contain enough material to give meaningful analytical data.

The following three examples from my own work give some idea of the size and form of biological specimens that can give valid quantitative x-ray microanalytical data.

9.7.3.1 Yeast Samples Encapsulated with Tea Tree Oil Containing ca. 1% Tetra-methyl Tin

The analysis was carried out on 4 µm diameter yeast cells which were quench frozen in melting iso-pentane at 113 K, stress fractured with a cold knife at 120 K, and sputter coated with 5 nm of platinum at 120 K. The specimen fracture faces were imaged and

226   9. Sample Stabilization to Preserve Chemical Identity

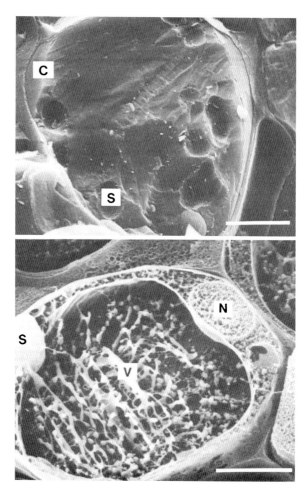

Figure 9.14. SEM images of a frozen-hydrated (upper) and frozen dried (lower) freeze-fractured root cells of the pond weed Lemna minor. The frozen-hydrated sample although relatively featureless shows the cell outline. The frozen-dried sample has increased topological contrast but reveals large voids once occupied by ice-crystals. C=cytoplasm; S=starch; N=nucleus. Sample coated with 8nm gold. Bar Marker = 10μm

analyzed at 120 K. Figure 9.15 shows an SEM image of this type of material. The SEM images were taken at magnifications of between x5–40 K, 5–10 keV, and 15–30 pA. A higher beam current of 50–100 pA was needed for the energy dispersive x-ray microanalysis. It was possible to identify and measure the major light elements found in biological specimens including large amounts of oxygen, together with a low concentration of tin.

9.7.3.2 The Location of Aluminum in Fresh Leaves of the Tea Plant

The analysis was carried out on frozen-hydrated fracture faces of different cultivars of young green Assam tea leaves. Three millimeter by 1 mm strips of leaf were cut, quench frozen in

Table 9.14. Retention of elements after different quench freezing and freeze substitution procedures

Element	Tissue	Substitution procedure	Percentage retained
N	Rat liver	Frozen in iso-pentane. Substituted in acetone 6–9 days at 194K	98
N	Rat liver	Frozen in iso-pentane. Substituted in methanol 6–9 days at 194K	95
P	Rat liver	Frozen in iso-pentane. Substituted in acetone 6–9 days at 194K.	98
P	Rat liver	Frozen in iso-pentane. Substituted in methanol 6–9 days at 194K	61
^{23}Na	Barley root	Frozen in iso-propane. Substituted methanol 6 days at 193K	1
^{23}Na	Barley root	Frozen in iso-propane. Substituted cetone 6 days at 193K	61
^{36}Cl	Barley root	Frozen in iso-propane. Substituted 20% acroline in ether 6 days at 193K	96

From Morgan et al., 1978.

Figure 9.15. Frozen hydrated image of brewers yeast at 120K. Living yeast cell culture microencapsulated with Tea Tree Oil containing alkyl tin, Cells quench cooled, fractured and sputter coated with 5nm of platinum at 120K. Small vacuoles at the cell periphery and nuclear material at the centre. Small amounts of tin found in the peripheral vacuoles. Picture courtesy Tony Burgess, Cambridge

slushy nitrogen at 63 K and quickly transferred to the pre-cooled specimen holder of an ancillary cold stage attached to the SEM. Cross-sections of the leaves were stress fractured with a cold knife at 130 K, sputter coated with 5 nm chromium at the same temperature and transferred to the pre-cooled cold stage of the SEM at 130 K.

Figure 9.16 shows a specimen image and the analysis was made on individual cells at between ×500 and ×5000 at 3 keV, 30 pA and 120 K. It was easy to identify the individual cells and their walls in the different tissues of the frozen leaves and energy dispersive analysis carried out on different regions of the leaf. The highest aluminum content was found in the cells walls of the two outermost regions of the leaf and high amounts of aluminum were associated with increased levels of magnesium and silicon. (P. Echlin, 2001).

9.7.3.3 Changes in Light Element Concentration During the Growth and Development of Tobacco Leaves

These studies centered on some earlier work carried out 25 years ago at the Philip Morris Research Center, Richmond, VA by the author and colleagues. Figure 9.17, from the paper by Echlin and Taylor (1986) on the x-ray microanalysis of fresh developing tobacco leaves, provides another example of low temperature x-ray microanalysis.

In all three examples, the specimens remained at or below the recrystallization of water throughout preparation, imaging, and

Figure 9.16. A low-temperature (120 K) secondary electron image of a frozen hydrated fracture face of a fresh young leaf of the tea plant *Camellia sinensis*. The leaf is composed of several distinct cell types and their individual cells can be separately analysed. The image was taken at 5 keV and 35 pA beam current. The magnification marker is 5 μm. (Echlin, unpublished image)

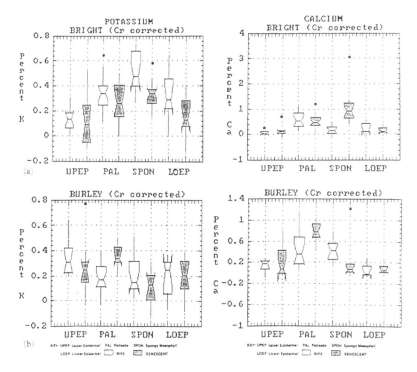

Figure 9.17. Elemental concentrations of potassium and calcium in four different cell regions of frozen hydrated Bright and Burley tobacco leaves at two stages of growth. The light figures are from ripe leaves, the dark figures are from later senescent leaves. The negative values are below the minimal detectable limit of 100 ppm. The Cr correction is for the spectral overlap of the Cr Lα line with the O Kα line.

Table 9.15. Relative peak to local back ground ratios of a number of elements in the same specimen of fresh green tobacco leaf mesophyll cells first in the frozen hydrated state and then after freeze drying

Hydration state	O	Na	Mg	P	S	K	Ca
Frozen-hydrated	52.0	0.06	0.04	0.29	0.26	0.52	0.09
Frozen-dried		0.04	0.13	0.28	0.37	0.47	0.72	0.56

analysis. In spite of the presence of ice crystals, they were smaller that the areas being measured by x-ray microanalysis. This is not a high resolution method but it provides quantitative analytical information at the cellular level of specimens.

Frozen hydrated samples are sensitive to beam damage and it is important to continually monitor the keV and beam current during analysis and as Table 9.15 earlier it is also important to maintain the samples either at a fully frozen hydrated or frozen dried state during analysis.

9.8 Summary of the Appropriate Stabilization Procedures for Preparing Samples for Chemical Analysis of Biological Specimens

The six general procedures listed below have been suggested as suitable stabilization methods for particular types of biological materials prior to chemical analysis in the SEM:

1. Ambient temperate procedures

    a. Wet chemical staining. Elements and small molecules
    b. Wet chemical histochemistry. Molecules
    c. Immunocytochemistry. Macromolecules

2. Low temperature procedures

    a. Frozen dried. Elements, molecules and macromolecules
    b. Freeze substituted Molecules and macromolecules
    c. Frozen hydrated. Elements

Each of these procedures have advantages and disadvantages and none of them alone will give the full information about the sample.

9.8.1 Wet Chemical Staining

This is only useful to reveal the general location and distribution of large groups of elements and molecules. For example, is there more or less calcium in bones than in skin? Is there more or less sodium chloride in seaweeds than in land plants? The disadvantage of this procedure is that it lacks spatial precision. Information obtained using backscattered electrons and x-ray photons.

9.8.2 Wet Chemical Histochemistry

This may be a little more specific than staining the sample and should provide some information about general groups of molecules such as proteins and carbohydrates. For example, which part of a seed contains starch? Where are the lipids, carbohydrate, and proteins in a section of animal skin? The disadvantage of this process is that it only makes distinctions between wide groups of chemicals. Information is obtained using backscattered electrons and x-ray photons. Immunocytochemical location: Provided care is taken during the initial stabilization this is an extraordinary sensitive and highly specific way to localize biologically active macromolecules in specimens. For example, location of arabinoxylanes is in the different tissues of cereal grains. Location of a specific cell detector is on the surface of a phagocyte. Information is obtained using secondary and backscattered electrons.

9.8.3 Freeze Drying

Although this method precludes the use of wet chemicals, it is not a good method when used alone. It could be used to locate heavy metals in a light element matrix provided the drying process had not disrupted the structural integrity of the specimen; for example, location of heavy metal contaminants in unicellular algae, single animal cells, and microorganisms. Freeze drying is more useful as a stabilization process *after* the dried samples are embedded in an appropriate resin. Information is obtained using backscattered electrons and x-ray photons.

9.8.4 Freeze Substitution

This is an important process when used in conjunction with immunocytochemistry. On its own, it has limited used as a stabilization process. Care has to be taken with the choice of wet chemicals used for initial stabilization, and this potential limitation makes it less useful as a method for localizing specific groups of elements and small molecules. The combination of freeze-substitution and immunocytochemistry has a wide application in living and functional biological specimens. Information obtained using backscattered electron.

9.8.5 Frozen Hydrated

This is the best stabilization procedure to use in association with the identification, spatial location, and quantification of elements from carbon to uranium (if necessary!) in biological material. The disadvantages of this approach is that the samples, by necessity, are small, are radiation sensitive, and need expensive additional instrumentation to ensure the specimen is rapidly quench cooled and then maintained below the recrystallization point of water. The information is obtained using x-ray photons.

9.9 Wet and Liquid Samples

The methods used for this type of sample are more or less the same as those used to stabilize and retain the structural integrity of these specimens. The central problem is to handle the liquids with minimal disturbance to their location and chemical integrity. There are three general approaches:

1. The wet and liquid samples may be untreated and chemically analyzed directly using either wet cells in the SEM or the variable pressure SEM.
2. The liquid and solid phases of the specimens may be separated by filtration, centrifugation, selective solubilization, sublimation, etc. and the two components studied separately.

The structure of the solid components is studied in the SEM, whereas the chemical components are best studied by standard analytical chemistry, which provides far more information than only using the SEM. The disadvantage of this approach is that it is no longer possible to retain the coincidence of structure and chemistry that is an essential feature of the SEM.
3. Samples such as the North Sea sandstone saturated with oil shown in the previous chapter may be quench cooled and the now solid immobilized liquid phase analyzed together with natural solid phase.

Although these three different approaches have been discussed in other parts of this book, there are a number of additional specialized methods that should be considered.

1. *Selective solubilization*. Oil bearing rock samples contain a mixture of organic and aqueous liquids. This liquid phase can be removed after impact crushing separation and the organic and aqueous components selectively dissolved in different solvents and analyzed separately by chemical means.
2. *Microdroplets*. One of the earliest uses of x-ray microanalysis in biology was to analyze small droplets of fluids taken from known sites in active biological material (Ingram and Hogben, 1967).

Using a high magnification binocular microscope, nanoliter volume samples are collected from various minimally perturbed tissue compartment using micropipettes. These are specialized techniques in which all sample handling procedures are carried out under paraffin oil to avoid drying and changes in the concentration of electrolytes in the droplet. A small volume (10–100 pl) of the liquid sample together with appropriate standards are placed on a polished beryllium support and freeze-dried. The miniscule piles of dried salts are analyzed by x-ray microanalysis using well established algorithms. This procedure goes some way to linking analysis and structure because the samples are taken from known compartment, i.e., specific glomeruli in the kidney. A good review of the techniques may be found in the paper by Quamme (1988) and the book by Warley (1997).

Morgan and Davies (1982) modified the microdroplet method so it could be used to extract larger sample from biological specimens. 1 ml samples taken from the specimen are sprayed as 2- to 10-µl microdroplets onto thin plastic films using a nebulizer. The droplets dry very quickly and their elemental content analyzed by x-ray microanalysis.

1. *Microincineration*. It has been known for some time that it is possible to remove water and organic material from wet biological samples by placing them either in a muffle-furnace at 773 K or by low temperature etching with a reactive oxygen plasma.

Although far from an ideal method, micro-incineration does, however, have a limited number of uses in microanalysis, provided it can be demonstrated that there is minimal and measurable dislocation of the ashed material. Albright and Lewis (1965) used the approach to detect Fe and Ca in bacteria. Davis and Morgan (1976) used the method as part of a preparation procedure for the analysis of tissue samples and found that it could be controlled sufficiently to retain some volatile elements, such as sulfur and chlorine. Microincineration also can be used to remove background organic material when the mineral content is low in sectioned tissues. Barnard and Thomas (1978) used the same approach, whereas when the x-ray sensitivity of sulfur, calcium, and phosphorous increased, there was a loss of chlorine. These and other studies indicate that this approach should only be applied to bound elements and elements of low volatility.

10

Sample Cleaning

This chapter is deliberately placed here in this book as it is the penultimate procedure we must consider during sample preparation. Once cleaned, the specimen either goes directly into the specimen chamber of a scrupulously clean SEM or via the charge elimination procedures discussed in Chapter 11. This is not to suggest that this is the only time cleaning occurs; it is a continuous and repetitive process during sample preparation to remove every thing that is not an original and integral part of the sample. If a particular preparative procedure contaminates the specimen, clean it off before going on to the next task.

The dictionary definition of the word clean is "free or relatively free of dirt or pollution" and "unadulterated or pure in various ways." The words "various" and "relatively" reveal that complete cleaning is an activity that is very difficult to achieve. For example, the *proper* cleaning of both sides of a domestic window is not an easy task.

The manufacturers of modern scanning electron microscopes go to great effort to ensure the interior of the microscope column is very clean. Field emission scanning electron microscopes use oil free ion-getter pumps and turbomolecular pumps to maintain the clean high vacuum inside the instrument. These pumps have replaced the older oil diffusion pumps which, as Figure 10.1 shows, can cause contamination in the microscope column.

Cleaning is particularly important in high spatial resolution imaging and in situ analysis of the nanostructures found in microelectronic materials, integrated circuits and semiconductors which are frequently examined at low voltages and beam currents. As users of the SEM, we must design, adopt and rigorously apply cleaning procedures to ensure the pristine state of the prepared sample is maintained.

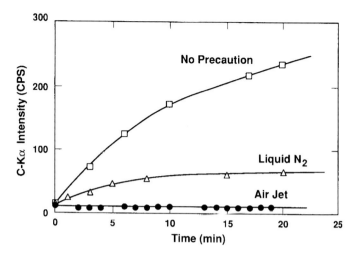

Figure 10.1. A polished copper sample showing a build up of carbon contamination in an oil diffusion pumped vacuum system in a SEM operating at 10keV and 100 nA. A liquid nitrogen anti-contaminator and an air jet reduced the levels (Goldstein et al., 2004)

Dirty specimens lessen the impact of the microscope in three ways.

1. Liquid, solid, and vapor contamination compromise the optimum operation of the SEM.
2. Contamination alters and degrades the physical appearance of the specimen.
3. Contamination changes and compromises the natural chemical properties of the specimen.

It is best to assume that all specimens are to some extent contaminated (and so too are our fingers). As a golden rule, always wear disposable plastic gloves and use clean metal tools, such as forceps or tweezers, when handling specimens, placing prepared specimens onto the microscope stage, and removing them after examination.

There are two general types of cleaning, non-contact cleaning in which there is no physical contact between the cleaning agent and the specimen, and contact cleaning in which there is physical and chemical contact between the cleaning agent and the sample. Ideally we would like only to use non-contact processes to ensure that the sample surface is undamaged. For example, loosely adherent dust can be removed from a dry specimen by a puff of clean air or, better still, a low pressure jet of an inert gas such as nitrogen. Most cleaning is achieved using solids, fluids, chemicals, or high energy beams that are applied to the sample surface by varying degrees of physical contact. These processes should not leave any traces of the cleaning materials as this may compromise any subsequent chemical analysis.

There are a large number of ways to clean a sample, and before setting out a particular cleaning protocol, it is important to first set out why the SEM is being used to study the specimen. For example, x-ray microanalysis of a rock sample should be carried out on a clean polished sample, whereas a structural analysis of the same sample is best performed on a clean natural surface. The following seven points require attention:

1. Is the sample to be examined for structural information, chemical information, or both?
2. Is any surface contamination derived from the sample itself, from its natural environment or as a consequence of the processes of handling the sample?
3. Is the sample to be studied at high or low magnification?
4. Assemble all the information available about the sample. For example is if soft, hard, inert, reactive, wet, dry, etc.?
5. Assemble all the information about how different cleaning procedures might have an adverse effect on the sample.
6. How stable will the sample be after it is cleaned?
7. How might the sample react to a chemical cleaning agent. For example, an acid wash will clean a metal surface, but will it destroy a biological specimen?

As a general rule the cleaning process should start by using the mildest cleaning agent and the least physical contact. There is only one rule; the cleaning process must not damage or modify the specimen. In the chapter on sample exposure, the methods considered included impact fracturing, cutting, sawing, grinding, polishing, abrasion plasma, and electrochemical etching. These are physically invasive and potentially damaging procedures and should not be used for cleaning specimens at the end of the whole sample preparation protocol.

10.1 Metals, Alloys, and Metallic Materials

Metallic samples are found either as alloys or metals in manufactured materials. They also are found in their natural state in geological samples as surface and internal components of both undamaged and broken specimens. Many such specimens can be examined with a minimum of sample preparation, although they must first be thoroughly cleaned. The cleaning process must remove mineral and organic oils, grease, and paint together with any traces of inorganic and organic chemicals on the surface. Figure 10.2 shows that one of the principal contaminants on metal samples is carbon and organic carbon compounds derived either from faulty cleaning or the microscope itself.

The cleaning of this group of specimens can be achieved in three ways:

1. Mechanical means such as high pressure particle abrasion such as sandblasting and lapping procedures that will certainly

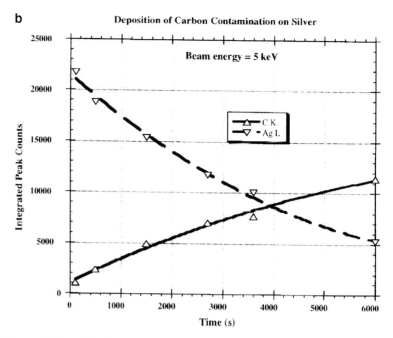

Figure 10.2. An EDS spectrum of a time sequence of intensity of the C Kα and Ag Lα lines from a silver specimen examined with a 5keV focuses probe scanning a 10^2 area on the sample. (Goldstein et al., 2004)

clean the sample but at the expense of surface erosion and damage. This approach should not be used as a final stage for topographical studies, but is a good preliminary cleaning process for preparing specimens for subsequent surface analysis.
2. Chemical processes such as powerful detergents, organic solvents, reactive acids, and alkalis. This approach can be used to clean metal surfaces for imaging in the SEM.
3. Electrolytic means such as electrochemistry and electrostripping can be used to prepare metal surfaces for both imaging and analysis.
4. The use of high-energy particle devices discussed in Chapter 6.

For topographic imaging, the surface must be degreased by washing in high purity solvents such as acetone, toluene, or alcohol, using an ultrasonic cleaner. It is important to use several changes of the degreasing agent. Plasma etching is a good way to remove both the surface hydrocarbons and those hidden in cracks, holes, and crevices. After drying the specimen, a jet of high pressure nitrogen can be used to remove any loose components. The same cleaning procedures can be used prior to the lapping procedures to prepare the highly polished metal surfaces required for quantitative x-ray microanalysis.

Table 10.1. A simple procedure for cleaning a metal surface

1. Remove adherent dry material with a tooth brush.
2. Remove oil and grease with a suitable solvent and a soft rag.
3. Treat metal surface with a suitable acid.
4. Dip into a boiling solution of ammonia and detergent solution and scrub well.
5. Rinse the metal surface ten times first in tap water and then distilled water.
6. Air dry the metal at ambient temperature.

Table 10.2. A more complicated method for cleaning a metal surface

1. Remove adherent dry material with a tooth brush.
2. Remove oil and grease with a suitable solvent and a soft rag.
3. Remove any oxide layers with an abrasive such as pumice.
4. Rinse well in running tap water.
5. Place the metal sample in a solution made up of 50g Na_2CO_3, 25g $Na_5O_{10}P_3$ and 25g NaOH in 2 litre of water.
6. With the sample as the cathode, electroclean for 1-3min in a current density of 1-3A/dm then reverse the polarity for 5-10 sec.
7. Rinse the metal surface ten times first in tap water and then distilled water.
8. Briefly rinse the metal surface in dilute sulphuric acid.
9. Rinse the metal surface ten times first in tape water and then distilled water.
10. The cleaned metal should be kept in running water to prevent oxidation.

There are a wide range of more specific procedures in which it is necessary to use a specific recipe to clean a particular metal surface without changing either its structure or natural chemical composition. An up-to-date version of *The Handbook of Chemistry and Physics* (CRC Press, Boca Raton, FL) is an invaluable guide to formulating a cleaning material recipe for a given metallic material. Cleaning solutions, accessories and equipment are available from many vendors, including Ernest Fullam Inc. (www.fullam.com), Agar Scientific (www.agarscientific.com), Spi Supplies (www.2spi.com), and Electron Microscopy Supplies (www.ems-diasum.com). Tables 10.1 and 10.2 provide general cleaning processes for metals.

10.2 Hard Dry Inorganic Materials

These samples are generally strong resistant materials that are chemically inert in their natural environment. It is usually necessary to cut the sample so that it fits into the microscope. The dust from the cutting process can be removed with a jet of nitrogen gas and non-porous samples cleaned by appropriate solvents such as distilled water, or acetone with or without, the use of a fine art squirrel-hair paint brush.

A more vigorous cleaning of non-porous specimens is achieved with an ultrasonic cleaner, in which the high frequency vibrations are transferred to the cleaning fluid producing a turbulent penetrating action. These cleaners are used in conjunction with water or solvents to remove contaminating material from crevices and small holes. Prolonged exposure to an ultrasonic cleaner may damage some softer specimens. A final rinse with methanol helps to remove any remaining surface films. In the earlier chapter on specimen exposure, a number of powerful physical techniques such as grinding, polishing, and abrading were described as appropriate ways to proceed with hard inorganic and metal specimens. These methods may be used to clean the specimen surfaces.

Hard inorganic materials such as bone, enamel, skeletal, and silicified structures are important component of biological specimen and most SEM studies are carried out on specimens in which both the soft and hard tissues remain intact and in situ. However, the hard materials are also of interest in their own right and bones, teeth and silicous deposits such as the phytoliths of grasses and the frustules of diatoms are important for specimen identification, classification and cladistics. The cleaning processes for these hard materials are draconian and involve a thorough removal of all the organic material without altering the inorganic structures. The soft biological components can be removed by washing in water or detergent solutions at low and high temperatures, dissolving in enzymes, or treatment with strong acids and bases. The cleaned samples must be thoroughly washed in distilled or de-ionized water. There are no half measures with these cleaning procedures and it is important to know that they do no damage the piece of hard material which is the focus of interest.

Hydrocarbons are one of the most persistent contaminants associated with hard inorganic materials. Figure 10.3 shows the build up of carbon contamination on a sample of different carbide and graphite. The contaminating hydrocarbons may initially be deposited during sample preparation and then subsequently undergo migration during sample examination in the electron beam.

As discussed, another source of contaminants comes from the microscope vacuum system. This is a potential problem during low voltage microscopy and for studies of semiconductors. Plasma cleaning is a very effective way of removing organic contamination before the sample goes into the microscope and Figure 10.4 shows a simple but effective device from XEI Scientific (www.evactron.com) to remove carbon from a silicon substrate.

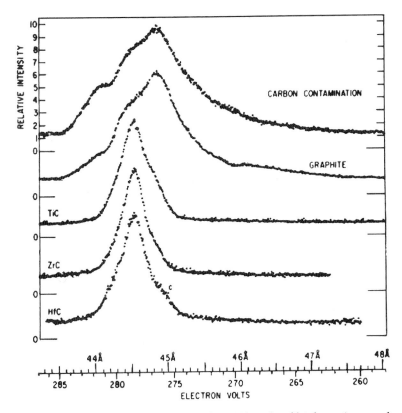

Figure 10.3. The graph shows the carbon K bands of high purity graphite and three different carbides covered by a contaminating layer of carbon after examination in the SEM at 4keV (Goldstein et al. 1981)

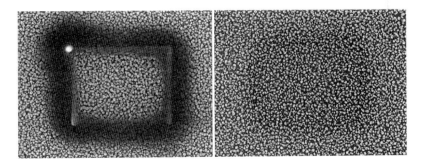

Figure 10.4. High resolution image (x50K) of a silicon "glass", a rough silicon substrate with a high density of etch pits. The left hand image is severely contaminated with carbon which is significantly decreased after 10 mins treatment with a XEI Scientific Evactron Anti-Contamination device (Picture courtesy XEI Scientific, Redwood City, California)

For very high resolution studies on microelectronic materials, the Fischione Automatic Sample Preparation System (ASaP, www.fischione.com) discussed at length in Chapter 6, is a very effective device to remove both organic and inorganic contaminants from specimens, as Figure 10.5 shows.

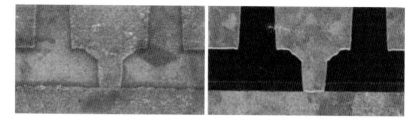

Figure 10.5. High resolution SEM images of a mechanically ground and oxidized cross section of a copper-based microelectronic material. The left hand image is of the material before cleaning. The right hand image is of the same sample following plasma cleaning, ion beam etching, reactive ion etching and ion beam sputter coating using the Fischione 1030 ASaP system (Picture courtesy E.S. Fischione Instruments, Export. Pennsylvania, USA)

10.3 Hard Dry Organic Specimens

These specimens are more chemically active that the two previous types of samples, and additional attention is needed with sample cleaning. Some of the specimens, such as paper and fine fabrics, are physically quite delicate and should only be cleaned by dry non-contact cleaning methods. Specimens of mature wood, wooden structures, and natural and artificial fabrics may be carefully washed either in water or a mild detergent solution with a final rinse in distilled water. The sample should be dried using clean, warm air. Jets of low pressure dry nitrogen remove small particles and a mini-vacuum cleaner can be used to remove dust from fabrics and paper provided there is minimum physical contact between the cleaning agents, and the sample should not automatically be used on the type of biological hard material sample discussed in the preceding unless it is possible to distinguish the point at which cleaning is complete and sample erosion begins.

10.4 Polymers and Plastics

These are physically more robust and chemically more inert than the organic samples of biological origin. In addition, most manufactured plastics and polymers are clean and probably need little more than spraying with a fine jet of water or a mild detergent solution followed by a final rinse with distilled water and drying in an ambient temperature, clean dust free environment. Unlike hard inorganic samples and metals, plastics and polymers can be damaged by organic materials, and care should be taken when contemplating any degreasing agents. Similarly, the plasma etching procedures that are so effective in removing hydrocarbons from rocks and metals should not be used to clean plastic and polymer surfaces.

10.5 Biological Organisms and Derived Materials

These specimens are the most difficult to clean properly because they are composed of a wide range of materials, and hence need a wide range of different cleaning protocols. There is a wide range of potential contaminants in and on biological specimens, and each sample has its own catalog of which parts are clean and which are contaminated. For this reason it is inappropriate to set out a precise guide to how to recognize contaminated samples in the SEM. However, Table 10.3 is a broad indication that the sample image may be contaminated.

With very delicate specimens it would be more appropriate to first stabilize the specimen before it is cleaned. Spurious surface materials such as mucus, secretions, body fluids, blood, cells, microorganisms, cell debris, mud, dust, etc., which frequently appear during sample exposure, must be completely removed before sample stabilization. The cleaning required has engendered a broad number of procedures and the general rule is to start off with the mildest cleaning agents and use the minimal physical contact.

Most soft specimens to be examined in the SEM for their structural features may usually be cleaned simply by gentle rinsing in an appropriate buffer at the correct pH, temperature, and osmotic strength. If the contamination persists, a dilute solution of a mild detergent can be used. The sample surface will be cleaned but will retain traces of the chemical remains of the buffer solution. The sample should then be stabilized and cleaned again a couple

Table 10.3. Some visual indicators of sample contamination occurring during the preparation of some polymers and most biological specimens

At the end of the stabilization procedures
The SEM image of the sample surface may show unexpected small bright or dark irregular precipitates. This may be due to contaminated preparative chemicals.

At the end of the dehydration procedure
The SEM image of the sample surface may similarly show groups of the same material seen at the end of the stabilization protocol. This may be due to contaminated, dehydration chemicals, and equipment.

After embedding
The SEM image of the sample may show smooth and rough surfaces which indicate incomplete resin polymerization. This is probably due to impure resin polymers.

During examination in the SEM
The image may show fine, regular dark and bright precipitates. This probably due to an exacerbation of the contamination during sample preparation and beam damage occurring during sample examination

of times with distilled water. For analytical studies it is careful to list the chemicals that have been used in the cleaning process.

For investigations of the surface features of small intact organisms and their discrete parts, from aquatic, marine, and damp terrestrial environments, the first action should be a gentle washing in several changes of one of the following solutions. If necessary, the small specimen can be collected by gentle centrifugation or one of the following liquids.

1. Ultrafiltered water from the aquatic or marine environment in which the organisms lives
2. An ultrafiltered balanced buffered salts solution that closely resembles the aquatic, marine, or wet terrestrial environment of the sample in respect to tonicity and pH
3. Ultrafiltered tap water

It may be possible to include a final rinse in de-ionized or distilled water provided there is an absolute assurance that the sample surface is not degraded.

These procedures should remove most of any adherent material on the outer parts of the sample. The only physical contact should be gentle brushing with a soft paint brush. Depending on the proposed imaging and analytical studies, it may be necessary to use more rigorous brushing, but ultrasonic cleaning should be avoided at this early stage. A slightly more vigorous approach is to add a drop or two of an appropriate surfactant, such as Tween-80, to the final washing fluids in order to increase the wettability.

The same approach may be used to clean the surface of large biological specimens although an added complication arises when it is necessary to cut out a piece of the specimen so that it will fit into the microscope. Internal cellular components may contaminate the surface. In these circumstances it may be necessary to prolong the washing and in extreme cases allow the cut sample to remain overnight in a gently agitated container of the appropriate cleaning fluid. The disadvantage of this approach is that it may well compromise any analytical studies.

Cleaning the interior of biological specimens is more complicated than cleaning just the natural surface. The mechanical processes used to reveal the inside will release the large number of the body fluids involved in respiration, digestion, excretion, reproduction, and growth. These materials, which consist of complex aqueous solutions and suspensions of carbohydrates, proteins, and fats, may be washed away using the procedures mentioned earlier, but some of them will need more rigorous attention. Buffered solutions of appropriate enzymes can be used to remove these three natural contaminants either individually or when combined, for example, in mucous, blood, sap, etc. Salts solutions and suspended material can be washed away by irrigation with one of the clean liquids mentioned earlier.

An alternate approach is to first flush out the interior liquid content of large specimens before dissecting them. The contents of the gastro-digestive system of vertebrates can be washed out by gentle lavage through both the anal and buccal openings. During dissection, it is impossible to completely avoid contaminating the internal spaces of the sample. Continuous gentle irrigation with balanced salt solutions and removal of the washing by aspiration will help to keep the internal spaces clean. By careful dissection, it is possible to remove some of the major organs such as liver, kidney, heart, etc. as intact structures. The contents of these large organs are a potential source of contamination. Blood and tissue fluids can be removed from the vascular tissue associated with the large organs by the process of perfusion or dripping described in Chapter 8.

It becomes progressively more difficult to apply these washing procedures on smaller and smaller animals. Plants present a different set of problems because of the very wide range of their water content, which ranges from 98% water in a young lettuce leaf to 2% in some seeds. In addition, very specialized conditions are needed to reveal the intact and functional internal structures of plants. The only safe ways to clean plant specimens is to either bathe the external structures of living plants with filtered tap water or use a gentle jet of clean air to blow away dust and detritus from dry plants. Bacteria and microscopic plants and animals may be cleaned by repeated gentle washing, precipitated by centrifugation and resuspended using an appropriate balanced salts solution.

10.6 Wet and Moist Samples

Cleaning this category of samples, which excludes biological material also represents quite a challenge. In the case of immiscible mixtures such as oil from marine deposits, it should be possible to dissolve the inorganic salts in distilled water, before freezing the oil and examining it at low temperatures. For liquids such as salad oil, there is sufficient difference in the density of the two main components of oil and water to separate out into two distinct layers. With moist materials such as clays, muds, pharmaceutical agents, paints, and many food materials it is difficult to see how such materials could be cleaned without severely altering both their structural and compositional features. The best one could hope to do is to physically separate the liquid and solid phases by filtration or centrifugation and attempt to wash the remaining solid phase with a clean, inactive liquid. It may be more appropriate to use analytical chemistry techniques such as mass spectrometry, spectroscopy, electrochemistry, and chromatography to analyze the liquid components of this type of sample.

11
Sample Surface Charge Elimination

11.1 Introduction

More than 60 years ago, Knoll (1941) discovered that a piece of mica examined in an early predecessors of the commercial scanning electron microscopes, appeared bright while the surrounding regions appeared dark. This phenomenon was referred to as charging and is a common feature of most secondary electron images of non-conducting specimens. This is because the secondary electrons are emitted with such low energies, 5–50 eV, that local potentials due to charging can have a large effect on the collection of the SE signal detector, which typically has a potential of +300 V. Table 11.1 shows the wide range of electrical resistivity of the six general types of specimens studied in the SEM together with some of the materials used as conductive coating layers.

Images of charging samples must not be confused with contamination, and the simple way to make this distinction is to turn off the primary beam and after 60 s, turn it on at a lower keV, and re-examine the same sample area. If the area occupied by the original higher magnification image is visible, then this is contamination rather than charging.

Charging gives rise to many weird and distorted images, but the simple expedience of emulating King Midas and covering the surface of the sample with a very thin layer of gold eliminates the problem. In the past 50 years there have been many advances in both SEM and x-ray microanalysis; gold coating is no longer always the best way to eliminate surface charging. This chapter gives details of the different ways it is now possible to solve this problem. At the outset it is important to appreciate that the methods described only serve to eliminate surface charging, and in nearly all cases the bulk of most specimens remain non-conductive (although metals are an obvious exception).

Table 11.1. Electrical resistivity of samples from each of the six different categories of specimens considered in this book. Resistivity expressed as Ω.cm at 300K

Metals	Resistivity	Hard dry inorganic	Resistivity
Magnesium alloy	9.0	Quartz	9×10^{10}
Manganese-steel alloy	70.0	Silicon oxide	7×10^{12}
Constantan (Cu/Ni)	49.0	Porcelain	1×10^{11}
Nichrome	10.0	Soda-lime glass	$1 \times 10^{10} - 10^{12}$
Hard dry organic		**Polymers**	
Wood	$5 \times 10^{12} - 10^{15}$	Polypropylene	1×10^{15}
Dry paper	$1 \times 10^{9} - 10^{10}$	Polyvinylchloride	$1 \times 10^{11} - 10^{14}$
		Cellulose acetate	$1 \times 10^{10} - 10^{13}$
Biological		**Liquid**	
Bone	5×10^{10}	Water	6×10^{7} to 10^{10}
Dried fruit	5×10^{8}	Ice	5×10^{11} (at 200K)
Common coating metals			
Carbon	3500		
Chromium	13.0		
Gold	2.40		
Palladium	11.0		
Platinum	10.0		

Although charging in the SEM is well studied, it still remains incompletely understood, and these fine details are not considered here. The following brief description helps to explain the phenomenon.

Consider the sample as an electrical junction into which a current i_B flows and from which backscattered (η) and secondary electrons (δ) are emitted. When a good conductor, such as a copper disc in good contact with ground, is struck by an incident beam i_B of 20 keV, the sum of i_η and i_δ accounts for 40% of the incident beam energy. The remaining beam energy flows to ground as specimen (or absorbed) current i_{SC}, which in the case of copper is 60% of the incident beam energy. The balance of the current may be expressed as Equation 11.1.

$$i_B = i_\eta + i_\delta + i_{SC} \qquad (11.1)$$

Only a negligible amount of the incident beam energy is used to generate x-ray photons and the visible photons used for cathodoluminescence. It is essential that the conductive path from the specimen to the ground is continuous during all phases of SEM imaging and x-ray microanalysis.

The lack of a continuous flow of the specimen current to ground becomes a critical problem when dealing with non-conducting specimens, which act as insulators. With the exception of most metals and metallic samples, hard dry inorganic and organic specimens, polymers, biological, and wet samples are insulators and invariably charge in the electron beam unless an artificial conductive pathway is introduced on or in the specimen to ground the specimen current.

For insulators, a region exists in which the number of emitted electrons, i.e., $i_\eta + i_\delta$, may be larger than the number of incident electrons i_B. This region is defined by two values of the incident beam E_1 and E_2 for which $i_\eta + i_\delta = 1$; these values are referred to as the first and second crossover points. E_1 is of the order of several hundred volts and E_2 ranges from 1 to 5 keV depending on the nature of the material. Table 11.2 gives the E_2 values for a number of different non-conducting materials.

If the incident beam energy is less than E_1 then $i_\eta + i_\delta$ is greater than 1 and fewer electrons leave the specimen than enter, resulting in a build up of a *negative* charge. The charge lowers the effective energy of the primary electron beam producing a further decrease in $\eta + \delta$. This situation continues until the specimen is charged to a sufficient level to totally deflect the incident beam of electrons.

If the incident beam energy is between E_1 and E_2, then more electrons leave the specimen than enter it, i.e., $i_\eta + i_\delta$ is less than 1. This charges the specimen *positively* and the buildup of the positive charge acts to decrease the effective value of δ since the low-energy secondary electrons are attracted back to the specimen. The effective value of $\eta + \delta$ becomes unity because of these procedures and

Table 11.2. The E2 values of different non-conductive hard dry inorganic materials, hard dry organic materials, polymers, and a volatile organic liquid

Material	E_2 (kV)
Resist	0.55
Resist on SiO$_2$	1.10
Pyrex glass	1.90
Chromium on glass	2.00
Gallium arsenide	2.60
Sapphire	2.00
Quartz	3.00
Alumina	4.20
Balsa wood	0.50
Teak wood	2.00
Kapton	0.40
Polysulphone	1.10
PBT Celanese	1.10
Nylon	1.20
Polystyrene	1.30
Exxon PC	1.30
Polyethylene	1.50
Rohn & Haas PMMA	1.60
Poly(vinyl)chloride	1.65
Teflon	1.80
Acetaldehyde-diethyl-acetal	1.65

Information courtesy of David Joy and Linda Sawyer.

a dynamic equilibrium is set up with the emitted current equal to the incident current in a small, positive, surface charge.

If the incident beam is greater than E_2, there is *negative* surface charging, which decreases the effective value of the incident beam, raising $i_\eta + i_\delta$ until an equilibrium is established with the incident beam. This equilibrium is unsatisfactory because of the variations in the final surface potential from place to place on the specimen and the slight leakage through surface conduction (Oatley, 1972).

For non-conductive specimens it is best to operate with $E_1 < E < E_2$ that for many specimens is achieved by setting the incident beam energy to 1 keV. An SEM with a field emission gun can obtain high resolution images at this sort of voltages.

Charging is manifest in many ways and most microscopists have their own catalog of odd, bizarre images. These perturbations range from annoying little flashes of light on the specimen surface, dark and light halos on sample images, grossly distorted images and the delightful image in Figure 11.1 shows how the incident electron beam images its own entry pathway into the specimen chamber.

A more typical charging artifact is sometimes seen when trying to record an image in the SEM. This is shown in Figure 11.2.

A similar list of artifacts occurs when attempting to carry out x-ray microanalysis on non-conductors. An example of one of these artifacts is shown in Figure 11.3.

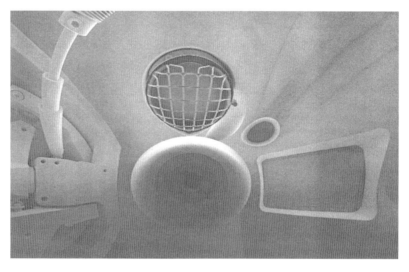

Figure 11.1. A view of the inside of the chamber of an SEM obtained using a highly charged uncoated clean glass sphere as a specimen. The charged surface acts as an electron mirror and enables one to see and identify the various components in the chamber. Picture courtesy Tony Burgess, Cambridge, UK

Figure 11.2. An SE image of a badly charging sample of an uncoated piece of a broken roof tile examined at 10 keV

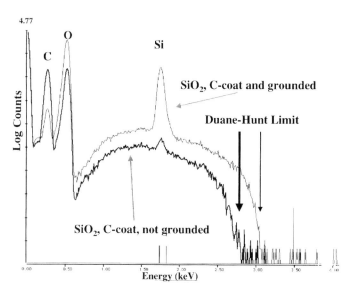

Figure 11.3. Detecting charging on an EDS spectrum. A specimen of silicon coated with 3 nm of carbon is studied with a 3 keV electron beam. The coated and grounded specimen has a Duane-Hunt limit of 3 keV. The coated but ungrounded specimen has a decreased Duane-Hunt limit of 2.5 keV and shows charging from 500 eV upwards. Picture Dale Newbury, National Institute of Standards and Technology, Gaithersburg, Maryland, USA

In addition to the image changes, the negative charge built up on a non-conducting specimen can both reduce and divert the incident beam away from the image, and the flux of generated x-rays is reduced and mislocated and appears to come from areas well away from the imaged area.

11.2 The Three General Ways to Obtain Optimal Image and Analytical Data from Non-conductive Specimens

1. To consider ways to reduce or eliminate surface charging only *after* the specimen is inside the microscope by making subtle changes to the way the instruments are operated
2. To consider ways to invasively and chemically modify the *whole sample* and make it totally conductive before it goes into the instrument
3. To consider ways to modify just the *sample surface* before they go into the instrument

11.2.1 Modifying the Microscope by Optimizing the Operating Conditions to Reduce Charging

As indicated in Chapter 1, it was decided to only consider the changes that can be made to the sample *before* it goes into the microscope. However, specimen charging is a unique phenomenon of the SEM and it would be erroneous not to briefly consider the changes that may be made to the operation of the microscope in order to reduce charging. Some of these changes have already been hinted at in the Introduction.

1. Lowering both the voltage and current of the incident beam diminishes charging. This is the first and easiest change to make.
2. Changing the scan speed from slow to TV rate diminishes the time the sample is exposed to the incoming beam.
3. Use the higher energy backscattered electron to image the specimen rather than the much lower energy secondary electrons, which are more readily influenced by both positive and negative charging.
4. Charging may be minimized for a given material by operating the microscope at a low voltage, with the E_1 selected to correspond to the E_2 upper crossover point as shown in Table 11.2. In the modern SEM, the accelerating voltage easily may be changed by small increments of 100 eV or less.

The E_2 values may be determined using the scan square test described by Joy and Joy (1996). The uncoated sample is placed in the SEM and imaged at ×100 at 3–5 Kev and a TV scan rate. The magnification is very quickly increased to ×1,000, maintained there for 5 s, and then immediately returned to ×100. The small scan area at the center of the screen is then examined.

1. If the scan area is *brighter* than the background, then the sample is charging *negatively* and the beam energy is greater than E_2 (or less than E_1).
2. If the scan area is *darker* than the background, then the sample is charging *positively* and the beam energy is less than E_2 (or greater than E_1).

Set the SEM at its lowest operating voltage and repeat the scan square test. If the sample is charging positively, then $E_1 < E_2$. Carefully increase the voltage and image the sample at the point where charging is minimized. If the sample is charging negatively, then $E_1 > E_2$ and it is not possible to lower the voltage any further, the sample should be tilted 45° and the scan square repeated.

1. When carrying out energy dispersive x-ray microanalysis, check that the maximum x-ray photon energy emitted from the specimen is equal to the energy of the incident electron beam producing the x-rays. This is referred to as the Duana-Hunt Limit.
2. Pre-bombard the sample with an argon beam that traps positive ions in porous surfaces, which suppresses the build-up of a negative charge.

The six procedures are non-invasive and easily performed on a modern SEM. The following three additional suggestions are minimally invasive procedures that can be performed *before* the sample goes into the microscope:

1. Impaling soft specimens on a mesh of fine conductive wire attached to the metal specimen holder.
2. Entrapping specimens in a low melting point conductive metal, i.e., gallium (303 K) or alloys such as Wood's metal (343 K).
3. Tightly wrapping samples in thin metal, such a clean aluminum baking foil, and cutting a small window in the foil to reveal the area of interest on the sample.

The latter three suggestions and other similar methods are a stopgap that can be tried out on precious and/or unique specimen such as museum, archaeological and fine art objects, forensic samples, and jewelry, which may not be changed in any way in order to obtain images and data.

11.2.2 Alter the Chemical Nature of the Whole Specimen and Convert It to a Conductor

Unlike the physical instrumental procedures described in the previous section, the effectiveness of the procedures to be discussed here depend on altering the chemical nature of the sample, primarily by marinating them in heavy metal salts. These methods are designed primarily for use with biological sample and for hard dry organic materials such as wood. Table 11.3 gives the

Table 11.3. Thermal conductivity and electrical resistance of some elements used for surface coating, for bulk conductivity staining and of some different types of non-conductive specimens examined in the SEM

	Thermal Conductivity (W/cm per °K at 300K)	Resistivity (Ohm/cm at 300K)
Elements used for surface coating		
Gold	3.17	2.40
Palladium	0.72	11.0
Platinum	0.72	10.0
Silver	4.29	1.60
Aluminium	2.37	2.83
Carbon	1.29	3500.0
Elements used for bulk conductivity		
Osmium	0.87	9.5
Lead	0.35	22.0
Uranium	0.28	30.0
Manganese	0.08	5.0
Some natural materials		
Marble	0.038	10^{14}
Polypropylene	0.001	10^{15}
Wood	0.05	10^{10}–10^{13}
Bone	0.05	10^{8}
Water	0.006	10^{4}–10^{7}

resistivity and thermal conductivity of some of the elements used to make samples more conductive.

11.2.2.1 Soft, Porous Biological Material

For biological specimens it is possible to combine ions of reactive heavy metals such as osmium, lead, manganese and uranium. Table 11.4 provides a list of such metals and some of their properties.

11.2.2.1.1 Reactive Heavy Metals
These heavy metals can be made to combine with bi-functional ligand mordents such as thiocarbohydrazide, phenylenediamine, and galloylglucosamine, which are usually referred to as tannins. Thiocarbohydrazide is a particularly useful mordent, as it can form a bridge between any osmium ions that have been used in specimen stabilization. One terminal amine group of the thiocarbohydrazide reacts with the osmium already in the specimen and the other amino group binds to further osmium ions, or to uranyl or lead ions.

Of all the mordents and heavy-metal ions which have been tried, the osmium–thiocarbohydrazide and osmium–tannin methods are the most successful and examples of these two procedures

Table 11.4. Thermal conductivity and electrical resistance of some elements which may be used for bulk conductivity staining

	Thermal Conductivity (W/cm per °K at 300 K)	Resistivity (Ω/cm at 300 K)
Chromium	0.94	13.0
Copper	4.01	1.67
Lead	0.35	22.0
Manganese	0.08	5.0
Osmium	0.87	9.5
Silver	4.29	1.60
Uranium	0.28	30.0
Zinc	1.16	5.92

are given in the following. These aqueous compounds can make a substantial increase in the metal loading in soft porous sample and significantly improve their conductivity. The bi-functional mordents act as molecular bridges between the sample and the heavy metal molecules. A good review of these procedures may be found in the papers of Murphy (1978) and Murakami et al. (1983). Two of the most successful bulk staining procedures for biological specimens are given in the following.

11.2.2.1.2 Osmium–Thiocarbohydrazide–Osmium Method (OTO)
1. Samples are first stabilized in a 2% glutaraldehyde solution in a 0.1 M cacodylate buffer at room temperature for 4–72 h.
2. Wash in buffer and transfer to a buffered 1% osmium tetroxide solution for 3–4 h at room temperature.
3. Thoroughly rinse the sample in buffer for 10 min.
4. Incubate the sample in a freshly prepared 1% aqueous thiocarbohydrazide solution for 10 min at room temperature.
5. Thoroughly rinse the sample in distilled water for 10 min.
6. Dehydrate, and if necessary, embed the sample for the SEM.

The times and temperatures for the initial stabilization and the OTO procedure can be varied depending on the porosity and size of the specimen. The thiocarbohydrazide–osmium sequence may be repeated several times (OTOTO---) in order to increase the loading of the specimen with the heavy metal. Figure 11.4 shows an example of a sample prepared by the OTO technique. Further details of this procedure can be found in the papers by Murakami et al. (1983) and Tanaka (1989).

11.2.2.1.3 Tannic Acid–Osmium Method (TAO)
1. Samples are first stabilized in a 2.5% glutaraldehyde solution in a 0.1 M phosphate buffer at room temperature for 6–8 h.
2. Continue stabilization in 6% glutaraldehyde in the same buffer.
3. Wash the material for 2 h in the same buffer.

256   11. Sample Surface Charge Elimination

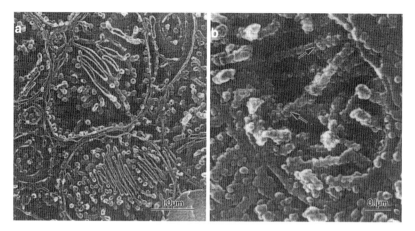

Figure 11.4. Bulk conductive stained rat mitochondria followed by a 1–2 nm layer of evaporated gold. Picture from Goldstein et al. (2004)

4. Rinse for 8 h in distilled water containing 2% tannic acid, 2% guanidine hydrochloride, 2% arginine, and 2% glycine.
5. Wash for 30 min in distilled water.
6. Stain the sample in 4% aqueous osmium tetroxide.
7. Wash for 30 min in distilled water.
8. Dehydrate, embed the sample for the SEM.

As with the OTO method, the times and temperatures of the initial stabilization and the TAO procedure should be varied depending on the porosity and size of the specimen. Tannin, or gallotannin, is a natural product derived from oak bark or nutgall and has a very complex chemistry. It is used as a mordent in dying fabrics and forms insoluble precipitates with albumin, starch, gelatin, and metallic ions. Further details of this procedure can be found in the papers by Murakami (1978) and Murakami et al. (1977).

Other bulk staining methods include tannic acid–osmium–thiocarbohydrazide (TAOTH), osmium–dimethylsulfoxide–osmium (ODO), silver nitrate-photographic fixer (SNP) and, rather surprisingly, osmium–hydrazine (OH). Hydrazine is a powerful reducing agent. It is very poisonous and a constituent of rocket fuel, so take care. Further details of these and other non-coating conductive staining methods may be found in the papers by Murphy (1978), Gamliel (1985a), and Tanaka (1989), and in Appendix 1 of the book by Postek et al. (1980).

The results of these bulk conductivity preparative techniques are impressive and high-resolution images have been obtained from completely uncoated samples. Images at magnifications as high as ×225,000 and a spatial resolution of 15–20 nm have been obtained from the tannic acid-osmium method at beam energies of 25 keV and 100 pA.

11.2.2.2 Hard, Dry Porous Organic Material

For hard dry organic materials it is possible to take advantage of the metallo-organic compounds used in some wood preservatives. Most of the 500 timber preservatives are organic but a few contain the elements copper and zinc. The UK Health and Safety Executive (www.hse.gov.uk) has a comprehensive listing of these chemicals, which are inordinately toxic and must be handled with great care and respect. Many wood preservatives are applied under pressure, and the time taken to infiltrate samples varies depending on the porosity and size of the specimen. Once the infiltration is complete, the sample should be dried to remove traces of organic material that might compromise the high vacuum of the SEM.

There are both advantages and disadvantages to bulk conductive staining techniques.

11.2.2.2.1 Advantages
1. Metal-impregnated samples may be repeatedly dissected to reveal additional internal structures without significant charging effects.
2. Very rough-textured or deeply indented surfaces can be rendered conductive.
3. The surface conductivity is more even and there are fewer edge effects on the images.
4. It makes it possible to obtain sufficient contrast from low-profile specimens.
5. Biological sample treated by these methods are mechanically more robust and show much less shrinkage during dehydration and drying.
6. It substantially increases the density of light element samples by the inclusion of heavy metal ions, the beam penetration is diminished, and the spatial resolution is increased.

11.2.2.2.2 Disadvantages
1. Sample preparation can take a long time.
2. The images have a somewhat lower signal to noise (S–N) ratio than specimens coated with a thin metal layer.
3. Many of the reagents and procedures can damage the samples and long exposure to osmium tetroxide mitigate against high resolution SEM at the molecular level.
4. The presence of large amounts of heavy metals precludes any meaningful attempt at x-ray microanalysis of light elements.
5. The same heavy metals make it virtually impossible to use immunocytochemical and histochemical procedures to analyze biological samples.
6. In some instances, the conductive staining procedures are so successful and result in a high BSE coefficient so that the topographic contrast is reduced.
7. The chemical procedures used in bulk staining are irreversible and may damage unique and precious specimens.

There is one last chemical approach to reducing charging in samples. Some success has been obtained by coating the sample surface with a fine spray of organic antistatic agents derived from polyamines, organic suspensions of noble metal colloids or graphite, or covering the sample with a thin polymer film of Formva or styrene-vinyl pyridine. These methods are only useful for very low magnification images and even then, the images are peppered with deposits of the dried spray particles.

The bulk conductivity procedures have a limited usefulness in a modern SEM with high spatial resolution at very low voltages and beam currents. The first approach to preventing charging is to carefully coat the sample surface with a thin conductive layer of metal. These procedures are discussed in the following section.

11.2.3 Modify Only the Surface Conductivity of the Sample Before Microscopy

The best and simplest way to overcome charging problems is to deposit (coat) a very thin (nm) layer of a conductive metal on the surface of the sample. Table 11.5 shows the thermal conduction and electrical resistivity of some elements and representative groups of specimens. The purpose of any film deposited on a sample is to provide a ground plane rather than to make the surface conductive. Any localized conductive region eliminates electric fields above the sample surface, and although the specimen is charging, incoming and outgoing electrons do not experience any fields.

The function of thin layers of conductive material is to:

1. Increase the surface electrical conductivity of the sample
2. Increase the surface thermal conductivity
3. In some cases, to increase the SE and BSA signal from the sample

Table 11.5. Comparative properties of some conductive elements and non-conductive specimens

Element/material	Thermal Conduction (W cm^{-1}/K^{-1} at 300K)	Electrical resistivity (μohm cm^{-1} at 300K)
Gold	3.17	2.40
Palladium	0.72	11.0
Platinum	0.72	10.0
Aluminium	2.37	2.83
Carbon	1.29	3500.0
Osmium	0.87	9.50
Lead	0.35	22.0
Uranium	0.28	30.00
Silver	4.29	1.60
Wood	0.05	$10^{10}-10^{13}$
Bone	0.05	10^{8}
Water	0.006	$10^{4}-10^{7}$

Before setting out to coat a sample for SEM imaging and subsequent x-ray microanalysis, the following questions should be considered:

1. What is likely to be the ultimate required spatial resolution of the images?
2. What elements are to be analyzed?
3. What are the x-ray peaks of the elements to be analyzed?
4. What is the total chemical composition of the specimen?
5. What is the form and nature of the sample?
6. What voltage and bean current is to be used for microscopy and analyzed?
7. Is the x-ray microanalysis to be carried out using an energy dispersive spectrometer or a wave-length dispersive spectrometer?
8. Is the x-ray microanalysis to be qualitative or quantitative?

Having obtained answers to these and other related questions, it is now possible to choose the coating material and how it should be applied in order to obtain the best images and most accurate analysis.

Industrial methods such as electroplating and anodization are not satisfactory because they rely on chemical solutions of high or low pH, which damage many samples. It is also difficult to deposit very thin layers of material by these methods. Glow discharge coating is not satisfactory because it essentially covers the sample with an uneven layer of impure soot. The only effective way of increasing the conductivity of a non-conducting sample is to coat the sample surface with a very thin conductive metal film.

A number of elements can be used to coat samples, and Table 11.6 gives an annotated list of their properties.

A film is defined as being between 0.5 and 10 nm thick. A great deal is known about how thin films are formed, and this information is useful when we come to consider the way SEM samples are coated. Figure 11.5 gives a diagrammatic representation of the steps in the formation of a thin film. The first atoms to arrive on the sample drift and diffuse on the surface, losing energy; eventually they reach a critical nucleation size where they begin to bind to each other and to the surface. The coating procedure should aim for many small nucleation centers that will quickly grow laterally, first to form islands and then a continuous film, with a minimum of discontinuous vertical growth.

Figure 11.6 provides a series of high resolution TEM images of the change during the growth of a thin film of silver that appears to form a continuous conductive layer at 6–7 nm.

The variables that affect thin-film formation include the binding and impact energy of the target material, the minimum temperature needed to evaporate the target, the minimum energy to sputter the target, the rate of metal deposition, the substrate temperature and binding energy, and the sample topography. Further details

Table 11.6. A list of elements which may be used to coat non conductive samples together applied to specimens. The section, Relative Sputtering Yield shows that not all elements Evaporation Techniques provides the following information. The elements (i.e. Ta, W) used of a nonmetallic refractory material, metal loops, coils, or baskets); whether the element For example, Gold has a good thermal conductivity, a low electrical resistivity and may

Element	Symbol	Thermal conductivity at 300 K (W cm^{-1} K^{-1})	Resistivity at 300 K ($\mu\Omega$ cm)	Melting point (K)	Boiling point (K)	Vaporization temperature (K) at 1.3 Pa
Aluminum	Al	2.37	2.83	932	2330	1273
Antimony	Sb	0.243	41.7	903	1713	951
Barium	Ba	0.184	50.0	990	1911	902
Beryllium	Be	2.00	4.57	1557	3243	1519
Bismuth	Bi	0.079	119.0	544	1773	971
Boron	B	0.270	1.8×10^{12}	2573	2823	1628
Cadmium	Cd	0.968	7.50	594	1040	537
Calcium	Ca	2.00	4.60	1083	1513	878
Carbon	C	1.29	3500	4073	4473	2954
Chromium	Cr	0.937	13.0	2173	2753	1478
Cobalt	Co	1.00	9.7	1751	3173	1922
Copper	Cu	4.01	1.67	1356	2609	1393
Germanium	Ge	0.599	89×10^{3}	1232	3123	1524
Gold	Au	3.17	2.40	1336	2873	1738
Indium	In	0.816	8.37	430	2273	1225
Iridium	Ir	1.47	6.10	2727	4773	2829
Iron	Fe	0.802	10.0	1811	3273	1720
Lead	Pb	0.353	22.0	601	1893	991
Magnesium	Mg	1.56	4.60	924	1380	716
Manganese	Mn	0.078	5.0	1517	2360	1253
Molybdenum	Mo	1.38	5.70	2893	3973	2806
Nickel	Ni	0.907	6.10	1725	3173	1783
Palladium	Pd	0.718	11.0	1823	3833	1839
Platinum	Pt	0.716	10.0	2028	4573	2363
Rhodium	Rh	1.50	4.69	2240	2773	2422
Silicon	Si	1.48	23×10^{4}	1683	2773	1615
Silver	Ag	4.29	1.60	1233	2223	1320
Strontium	Sr	0.353	23.0	1044	1657	822
Tantalum	Ta	0.575	13.1	3269	4373	3273
Thorium	Th	0.540	18.0	2100	4473	2469
Tin	Sn	0.666	11.4	505	2610	1462
Titanium	Ti	0.219	42.0	2000	3273	1819
Tungsten	W	1.74	5.50	3669	6173	3582
Vanadium	V	0.307	18.2	1970	3673	2161
Zinc	Zn	1.16	5.92	693	1180	616
Zirconium	Zr	0.21	40.0	2125	4650	2284

11.2 The Three General Ways to Obtain Optimal Image

with some of their properties. The list also includes details of how these elements may be may be sputtered using the equipment commonly available in a laboratory The section, to construct the holder of the coating material; the form of the element holder (i.e. a crucible being coated alloys or wets the material used to hold the element.
be evaporated from a W wire basket or may be sputter coated together.

	Latent heat of evaporation (kJ g^{-1})	Specific heat at 300 K (J g^{-1} K^{-1})	Relative sputtering yield (atoms/ 600 cV Ar+)	Evaporation technique
Al	12.77	0.900	1.214	Ta, W; Coil or Basket; Wets and alloys.
Sb	1.603	0.205	NS	Ta; basket, wets
Ba	—	0.19	NS	Ta, Mo, W; basket, boats; wets
Be	24.77	1.825	NS	Ta, W, Mo; basket; wets, toxic
Bi	0.855	0.124	NS	Ta, W, Mo; basket, boat
B	34.70	1.026	NS	Carbon crucible, forms carbides
Cd	1.199	0.232	NS	Ta, W, Mo; basket, boat, wets, toxic
Ca	—	0.65	NS	W; basket
C	—	0.712	NS	Pointed rods, resistive, evaporation
Cr	6.170	0.448	1.3	W; basket
Co	6.280	0.456	1.4	W; basket, alloys easily
Cu	4.810	0.385	2.3	Ta, W, Mo; basket, boats, not wet easily
Ge	—	0.322	1.2	Ta, Mo, W; basket, boat, wets Ta, Mo
Au	1.740	0.129	2.8	Mo, W, loop; basket, boat, wets, alloysTa
In	2.030	0.234	NS	W, Fe; basket; Mo, boat
Ir	3.310	0.133	NS	Refractory; thick W loop
Fe	6.342	0.444	1.3	W; loop, coil, alloys easily
Pb	0.857	0.159	NS	Fe; basket, boat, not wet W, Ta, Mo
Mg	5.597	1.017	NS	Ta, W, Mo; basket, boat, wets
Mm	4.092	0.481	NS	Ta, W, Mo, loop; basket, wets
Mo	5.115	0.251	0.9	Refractory
Ni	6.225	0.444	1.5	W; coil (heavy), alloys
Pb	0.370	0.245	2.4	W; loop, coil
Pt	2.620	0.133	1.6	Stranded with W, Ta; alloys
Rh	4.814	0.244	1.5	Sublimation from W, low pressure
Si	10.59	0.703	0.5	BeO crucible, contaminates with SiO
Ag	2.330	0.236	3.4	Ta, W, Mo; coil, basket, boat; not wet easily
Sr	—	0.298	NS	Ta, W, Mo; basket; wets
Ta	4.165	0.140	0.6	Refractory
Th	2.340	0.133	NS	W; basket, wets
Sn	2.400	0.222	NS	Ta, Mo; basket, boat; wets
Ti	9.837	0.523	0.6	Ta, W; loop, cool, basket; alloys
W	4.345	0.133	0.6	Refractory
V	9.000	0.486	NS	W, Mo; basket; alloys with W
Zn	1.756	0.389	NS	Ta, W, Mo; loop, basket; wets
Zr	5.693	0.281	NS	Ta; basket; forms oxide

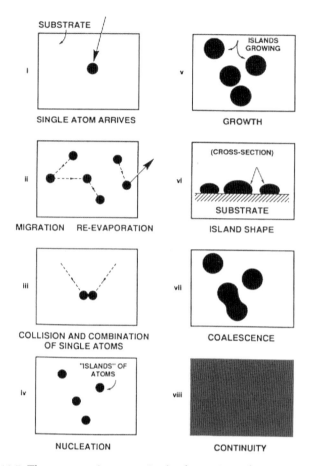

Figure 11.5. The progressive steps in the formation of a continuous thin film.

of thin-film formation can be found in the books by Maisell and Glang (1970) and Smith (1995).

There are two procedures for applying a thin coating layer under vacuum:

1. The evaporation method by which a conductive metal is heated to its vaporization temperature in a high vacuum and the evaporated metal atoms condense on the specimen's surface.
2. The sputtering method by which a negatively charged conductive metal surface is bombarded with positive ions in a low-vacuum gas discharge and the eroded metal atoms condense on to the specimen surface. These two methods are discussed in the following.

11.3 Surface Coating by Vacuum Evaporation

Most metals, when heated in a vacuum, evaporate once their temperature is raised to the point at which their vapor pressure

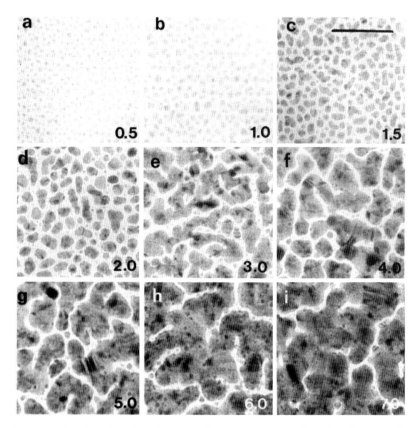

Figure 11.6. A series of high-resolution TEM images showing the growth of a silver film at 293 K. After nucleation, the crystals coalesce laterally (a–e), and then the smaller crystals grown on top of the existing crystals (f–h) and the gaps are progressively filled in to form a continuous layer (i). The numbers on each image is the calculated film thicknesses in nm

exceeds 1 Pa (~10^{-2} Torr). Many of the materials that are used for evaporative coating have a high melting point and need special procedures.

Refractory metals such as tungsten, tantalum and iridium must be heated in a refractory crucible by means of an electron beam. A paper by Ryazantsev (2000) describes a devise called the "Mini-Electron Gun," which incorporates an electron beam evaporator and is used to evaporate tungsten at a rate of 0.6–1.0 nm/min. The equipment also may be used to evaporate platinum-iridium and tungsten-tantalum. The system operates at 3–5 keV and 10–30 mA, and although designed to shadow-coat specimen for examination in the TEM, samples may be rotated during coating and could be used to apply a high melting heavy metal to specimens to be examined in the SEM.

Lower melting point materials such as the noble metals, are placed either in a refractory metal container or draped over a refractory metal wire which is resistance heated. In all cases, the material is evaporated uni-directionally in the same way as a light source illuminates an object. Vacuum evaporation may be carried out at high or low vacuum.

11.3.1 High-Vacuum Evaporation Methods

High-vacuum evaporation is one of the most popular ways of applying thin-metal and carbon films as coating layers on non-conducting specimens. Figure 11.7 is a diagram of the main features of a high-vacuum evaporation unit.

A number of different companies make vacuum evaporation coaters. Figure 11.8 is a desk top model available from Quorum Technologies (www.quorumtech.com) that is suitable for coating specimens for both SEM and x-ray microanalysis.

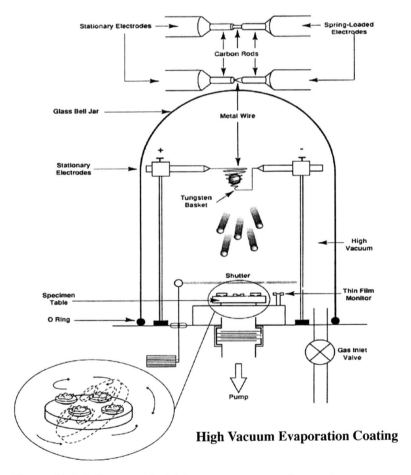

Figure 11.7. A diagram of a high vacuum evaporation coater

Figure 11.8. A high vacuum evaporation coater suitable for producing thin conductive coating layers on specimens to be examined in the SEM. Picture courtesy QuorumTechnologies Ringmer, Sussex, UK

11.3.1.1 Choice of Coating Material

The noble metals and their alloys, i.e., gold, platinum and a 4:1 gold:palladium alloy, are highly conductive and are suitable as a coating material for most samples to be imaged using electrons. The gold:palladium alloy is preferable to pure gold because the evaporated gold has a large grain size when viewed at high resolution. The advantage of using an alloy rather than a pure metal is that as the two different atomic species land on the specimen's surface; the presence of one discourages epitaxial growth of the other and thus increases the probability of obtaining a near-featureless thin film. Platinum films have a smaller grain size than either gold or gold palladium. Table 11.7 gives some of the properties of these noble metals.

If the sample is to be imaged using backscattered electrons, high density metals such as iridium, platinum, and tungsten may be considered, although the conductivity of iridium and tungsten is not as good as the noble metals. Table 11.8 provides some properties of these materials. Although silver has a high electrical

Table 11.7. Physical properties of three elements which are used to coat non-conductive materials for routine imaging in the SEM

Property	Gold	Platinum	Palladium
Density (Kgm^{-3})	19300	21450	12020
T-Conduction (W.cm)	3.17	0.716	0.718
Resistivity (µohm.cm)	2.40	10.0	11.0
Evaporation-T (°C)	1738	2363	1839
Sputter yield at 600eV	2.8	1.6	2.4

Table 11.8. Physical properties of some elements which may be used to coat specimen for high spatial resolution SEM

Property	Iridium	Tantalum	Tungsten	Chromium
Density (Kgm^3)	22421	16000	19350	7200
T-Conduction (W.cm)	1.47	0.575	1.74	0.93
Resistivity (µohm.cm)	6.10	13.1	5.50	13.0
Evaporation-T (°C)	2829	3273	3582	1177
Sputter yield at 600eV	1.2	0.6	0.6	1.3

conductivity, it is not used as a coating material because the thin metal surface quickly oxidizes and loses its conductivity.

Carbon, although a relatively poor conductor, is one of the best coating materials for studies involving x-ray microanalysis, provided one of the elements of interest is not carbon. For special imaging at high resolution, it is best to use a platinum–carbon or a platinum–iridium–carbon mixture as the evaporated material gives a coating material virtually devoid of any structure when viewed at high magnification (Wepf et al., 1991).

It is possible to evaporate a wide range of metallic target materials, and the information in Table 11.6 should be consulted when choosing the more exotic coating processes. It is important to only use high purity material. Although the evaporation procedures are more time-consuming than sputter coating, it can be used as a routine procedure for coating samples for both SEM and x-ray microanalysis. It is an excellent method for producing the very thin coating layers with very small grain size needed for high-resolution studies.

11.3.1.2 Preparing the Equipment

A generic protocol for carrying out high-vacuum evaporation is given in the following, although it is important to also follow the operating instructions given by a particular manufacturer. It is important that the bell jar, evaporation source, and coating platform are clean. Disposable gloves always should be worn when cleaning and setting up the evaporation chamber and when handling anything that is going to be exposed to the high vacuum inside the bell jar.

11.3.1.2.1 Check That the Inside of the Bell Jar Is Clean
The interior of the glass bell jar should be quickly wiped clean with a piece of soft paper tissue each time after it is used. Occasionally, or if it appears necessary, clean any metal parts with a detergent solution in an ultrasonic bath, rinse well in distilled water, air dry, and finally wipe over the surfaces with methanol and allow to dry. If the bell jar is visibly contaminated with deposited metal, it should be washed with a mild detergent solution, rinsed with distilled water, dried, and wiped with methanol. Abrasive cleaners should not be used on the bell jar.

11.3.1.2.2 Check That the Vacuum Pump Lines and Dry Nitrogen Gas Lines Are Secure
The bell jar seals should be clean and free of fine fibers and particles. With modern evaporation units there should be no need to use vacuum grease on the bell jar seal every time the system is used. If necessary, wash the bell jar elastomer seal with a mild detergent solution, rinse, air dry, and wipe with methanol. If deemed necessary, apply a *very thin layer* of an appropriate vacuum grease that is then wiped away using a lint-free cloth.

11.3.1.2.3 Assemble the Coating Source and the Specimens to Be Coated
Using disposable gloves, assemble the evaporation source and/or carbon electrodes. Check that the specimen table rotates and tilts, that the shutter above the specimen table is in position and the thin film monitor is operational. The clean samples should then be located at identifiable sites on the specimen table.

11.3.1.2.4 Pump the System Out to Its Working Pressure
This should be carried out using the mechanical and diffusion pumps, or better still, a turbo-molecular pump. Experience will show how long this process should take; significantly longer pumping times usually indicate that the bell jar seal is leaking and/or contaminated and needs cleaning. During the initial pump down, it is advisable to check the continuity of the electrical links to the evaporation sources, specimen rotation device and the thin-film monitor. Once the working vacuum of 10^{-3} and 10^{-5} Pa (~10^{-5} – 10^{-7} Torr) has been reached, any anti-contamination devices may be filled with liquid nitrogen.

11.3.1.2.5 Final Cleaning of the Evaporation Source
Make sure that the shutter is in place over the samples. The evaporation source should be gently heated to a cherry-red color to ensure any residual contaminants are vaporized. The high vacuum gauges usually show a small increase in pressure during this out-gassing.

11.3.1.3 Coating the Specimen

1. Turn on the specimen rotation device, and set the thin film monitor to the required value. While observing the evaporation source through welder's glasses, gradually increase the power to the electrodes until the metal melts and appears to shimmer. Open the shutter between the samples and the evaporation source and increase the power slightly to allow the molten metal to evaporate. Additional care should be taken during this final stage in order to prevent the now molten coating metal from falling off its refractory support and being splattered over the carefully prepared sample. If at all possible, evaporate upward.
2. The same procedure should be followed for carbon evaporation, except that in the final stages the power should either be pulsed or applied as a single flash in order to diminish the heat load onto the specimen. The coating for both metals and carbon is usually over in a few seconds. The required coating thickness is discussed later in this chapter.

11.3.1.4 Final Stages of Vacuum Evaporation

Turn the power off once the required thickness of material has been deposited and then closes the shutter and turn off the specimen table rotation. Allow a couple of minutes for the electrodes to cool off and then turn off the pumping system and slowly back fill the bell jar to atmospheric pressure with clean dry nitrogen. The specimens should be removed using forceps and/or disposable gloves and placed in a specimen box, which in turn should be placed in a desiccator.

11.3.1.5 Some Problems Associated with Evaporative Coating

1. Metal alloys do not evaporate congruently. This probably is not a problem with gold–palladium alloy because the two evaporation temperatures are only 100° apart. The evaporation temperatures of the iridium–platinum–carbon mixture differ by 600°, which may lead to particle separation.
2. The evaporated material travels in straight lines and may not produce an even and continuous layer on specimens with a rough or uneven surface. This problem may be overcome by tilting and rotating the sample in a planetary motion during the coating process.
3. The high temperatures needed to evaporate most metals may cause thermal damage to delicate polymers and biological samples. This problem may be alleviated either by placing a rapidly rotating shutter between the source and the sample or pulsing the power to the evaporation source.

11.3.2 Low-Vacuum Evaporation Methods

Although vacuum evaporation coating methods customary operate at a pressure of between 10^{-3} and 10^{-5} Pa (~10^{-5}–10^{-7} Torr) it is possible to evaporate carbon at 1 Pa (10^{-2} Torr) to produce carbon substrates and carbon thin films. Figure 11.9 is a bench top carbon coater available from Ted Pella Inc. (www.tedpella.com).

1. Disposable gloves should be worn when cleaning, setting up the carbon electrodes in the low vacuum chamber and placing the sample on the specimen table.
2. The two carbon electrode are assembled or high purity carbon string is fitted between the two electrodes and the sample is placed on the specimen table. The system is first pumped down to a high vacuum of between 10^{-3} and 10^{-5} Pa (~10^{-5} – 10^{-7} Torr) and then backfilled several times with dry argon or nitrogen and then pumped down to 1 Pa (10^{-2} Torr) and maintained at this pressure by adjusting the dry inert gas inlet valve.
3. The electrodes should be first slowly heated to a cherry red color to allow any organic contaminations to outgas. The carbon is then quickly evaporated resistively while the specimen rotates and tilts on the specimen stage. The evaporated carbon atoms have a short mean-free pathway in the poor vacuum. As they collide with the inert gas atoms, they scatter and provide an effective multidirectional coating layer.
4. Once the evaporation is completed, turn the power off and stop the specimen table rotation. Allow a couple of minutes for the electrodes to cool off and then turn off the pumping

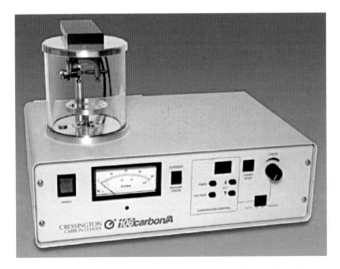

Figure 11.9. An Auto Carbon Coater for producing thin films on non-conductive SEM specimens for x-ray microanalysis. Picture courtesy Ted Pella Inc, Redding, California, USA

system and slowly back fill the bell jar to atmospheric pressure with the clean dry inert gas.

The advantage of the low-vacuum carbon evaporation method is that it provides a better coating layer on rough samples and those to be studied by x-ray microanalysis. The disadvantages are that the procedure is prone to contamination, and the high evaporation temperature of carbon (~3,000 K) may damage heat sensitive samples. The low-vacuum evaporation method, however, will not produce high-resolution thin films.

11.4 Surface Coating by Sputtering

Sputter coating is a popular and relatively simple way to apply a thin layer of metals and their alloys to non-conductive substrates. The target material is exposed to an energized plasma formed from a heavy inert gas such as argon. The target surface is eroded by the plasma, and target atoms are ejected and collide with the residual gas molecules. Target atoms have a very short mean free path and collide with each other and provide multidirectional coating on a stationary specimen.

The four main factors that influence the rate of sputtering are voltage, plasma current, target material, and nature of the inert gas:

1. The voltage influences the energy of the bombarding inert gas.
2. The plasma current has a direct effect on the rate of sputtering. High sputtering rates are obtained at high plasma energy.
3. The binding energy of the target material has an important influence on the rate at which it erodes. For example, the binding energy of gold and palladium is relatively low and will sputter much faster than tungsten, which has a high binding energy.
4. The higher the atomic number of the bombarding inert gas, the faster it will erode the target. Thus xenon gas plasma gives a higher sputtering rate than argon, which in turn gives a higher sputtering rate than nitrogen. Most sputter coaters use argon.

These four factors are variable and may be altered for different coating applications. Although high sputtering rates shorten the time taken to coat a specimen; there is an increased chance of thermal damage and an increase in the grain size within the sputtered film. There are several different ways to sputter coat materials, including the use of plasma magnetron coaters and ion beam and penning sputter coaters.

11.4.1 Plasma Magnetron Sputter Coating

Figure 11.10 shows a diagram of the main features of a modern plasma magnetron sputter coaters, which are readily available from many different suppliers. The interchangeable thin-foil metal target is the cathode and the specimen is the anode.

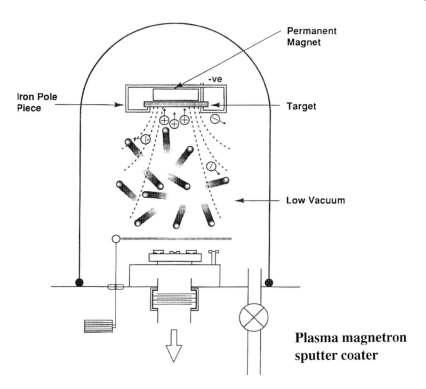

Figure 11.10. Diagram of a plasma magnetron sputter coater

Goodfellow Metals (www.goodfellow.com) are an excellent source of different high purity target materials that can be cut to size. The chamber is evacuated and backfilled with a high purity gas at low pressure and a DC voltage of 1–2 keV is applied to form gas plasma. A permanent rare earth magnet behind the target focuses the gas plasma on to the target to ensure a high sputtering rate and deflects electron away from the sample. The sample can, if necessary, be placed on a Peltier-cooled stage.

There are a number of different plasma magnetron sputter coaters on the market. Figure 11.11 shows the desk top sputter coater from Quorum Technologies.

Although one should follow the maker's instructions when using the equipment, the following general principles may be useful. The inside of the coater and the vacuum seals should be kept clean. Figure 11.12 shows the inside of the bell jar and the specimen coating area of a very dirty sputter coater contaminated with gold. The target area was similarly contaminated. A buildup of metal material may give rise to electrical conductivity problems and, if the sputter coater is uses with different targets, there may be doubts about the chemical composition of the ensuing thin film. The easiest way to prevent such a contaminating layer is to quickly wipe clean the bell jar and specimen coating region with

11. Sample Surface Charge Elimination

Figure 11.11. A compact turbomolecular pumped sputter coater suitable for preparing thin films for coating specimen for SEM. Picture courtesy Quorum Technologies, Ringmer, Sussex, UK

Figure 11.12. A view of the inside of a badly contaminated sputter coater. There has been no attempt to either clean the bell jar or the surface on which the specimen holders are placed

a clean, soft tissue after each coating sequence. Normally it is not necessary to clean the target material.

Plasma magnetron coaters always should use high purity argon as the inert gas, which should be connect to the coater with a stainless steel connection and metal-to-metal seals, if possible. Never use nitrogen to form the plasma as its sputtering rate is much slower than argon. *Never* use air, because the active plasma breaks down the oxygen and carbon dioxide into reactive components that will damage the sample. For high-resolution studies and low temperature microscopy and analysis, the coater should be evacuated using a turbo-molecular pump. Some coaters are evacuated with a rotary pump, and it is important to install an activated alumina trap to avoid oil vapor back streaming between the pump and the chamber. A generic protocol for plasma magnetron sputter coating is given in the following.

11.4.1.1 Choice of Coating Material

Although a wide range of metal targets may be used in plasma magnetron sputtering, a target of a 4:1 gold:palladium alloy is the best choice for routine imaging in the SEM because epitaxial growth of the deposited film is limited. Pure gold targets may be used for low (~×5–10 K) magnifications, gold–palladium for medium (~×10–50 K) magnifications and platinum targets for high magnifications above ~ ×50 K. All the noble metals have a relatively high sputtering rate at quite low plasma currents.

For very high-resolution studies, in which the coating layer must have a negligible thickness and very small particle size, it is necessary to use platinum, chromium, or one of the refractory metals such as tungsten or tantalum. A careful study by Stokroos et al. (1998) has shown that very thin layers of gold–palladium coating were a good alternative to either chromium or platinum for the application of the FESEM to biological tissues because of its higher yield of secondary electrons. However, this finding is strongly dependent on the type of sputtering device employed. The sputtering rate of refractory metals is much slower than the noble metals, and it is necessary to use high plasma currents and/or much longer coating times. These longer times increase the risk of damaging thermolabile samples and it is advisable to pulse the coating process.

It is possible to sputter both chromium and aluminum using a high efficiency sputter head and very pure argon. Both metals quickly oxidize, and a lot of energy is needed to break the oxygen–metal bond. The sample should first be shielded and the oxidized layer on the target surface removed using high plasma current. The shield is then removed and the metal coating can proceed at relatively low plasma currents. The coated sample should be transferred to the SEM as soon as possible. Figure 11.13 is a comparison between a specimen of biological material plasma sputter coated with very thin layers platinum or chromium.

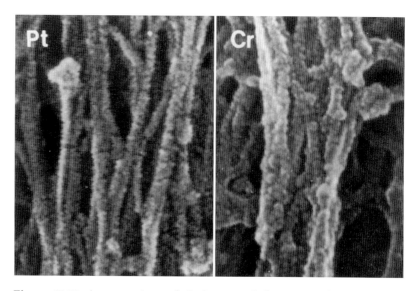

Figure 11.13. A comparison of platinum and chromium plasma sputter coated cytoskeleton from an extracted fibroblast. The surface regularities along the length of the microfilaments are partially obscured on the platinum coated sample. This is not the case with the chromium coated material. The coating thickness of both samples is 1.5–2.0 nm mass thickness, Image width = 137 nm. Picture from Goldstein et al. (2004)

Carbon usually cannot be sputtered with a laboratory plasma magnetron coater. The carbon–carbon bond energy is high, and it would be necessary to use high plasma energies to break these bonds. In addition, the sputtering would need to be carried out in a scrupulously anoxic environment. There are claims by some companies that their rather simple sputter coaters can sputter carbon. These assertions must be balanced against the assurance that there is no contamination in the system.

11.4.1.2 Preparing the Equipment

A generic protocol for carrying out sputter coating is given in the following, but it is important to also follow the operating instructions given by a particular manufacturer. It is important that the sputter coater is clean, and disposable gloves should always be worn when cleaning and setting up the target metal and loading the specimens on to the sample table.

1. Check that the inside of the bell jar is clean. The interior of the glass bell jar should be quickly wiped clean with a piece of soft paper after each time it is used. If necessary, clean any metal parts with a detergent solution in an ultrasonic bath, rinse well in distilled water, air dry and finally wipe over the surfaces with methanol and allow to dry. If the bell jar is visibly contaminated with deposited metal, it should be washed with a mild detergent

solution, rinsed with distilled water, dried, and wiped with methanol. Abrasive cleaners should not be used on the bell jar.
2. Check that the vacuum pump lines and argon gas lines are secure.
3. Loading the sample into the sputter coater.

This should be done using disposable gloves and/or the specific tools that are supplied with the sputter coater. The underside of each of the sample specimen stubs should be labeled, because nearly all sputter coaters can accommodate at least six specimens at a time. The clean, out-gassed samples are loaded onto the specimen platform and the system pumped out to a vacuum of no more than 1.3 Pa (10^{-2} Torr) At lower pressures, there is an increase chance of oil back streaming from the rotary pump.

11.4.1.3 Coating the Specimen

11.4.1.3.1 Flushing the Bell Jar
With the vacuum pump running, the chamber should be flushed and re-pumped at least five times. High purity argon is allowed into the system by opening the inlet needle valve and allowing the pressure in the chamber to rise to 65 Pa (5×10^{-1} Torr) for about 10 s, after which the needle valve is closed and the process repeated. Once the system is thoroughly flushed, the sample may be cooled if necessary. Some target materials, such as chromium, may form a surface oxide layer. This can be removed by pre-etching the target, making sure there is a shutter between the sample and the target.

11.4.1.3.2 Sputter Coating the Specimen
The argon gas inlet needle valve is gently opened by about half a turn and the pressure inside the bell jar allowed to fall to 13 Pa (10^{-1} Torr). The high voltage supply turned on and set to between 500–1,500 V at which point a light blue-colored gas plasma will form in the region of the target. The needle valve should be slowly opened further until a plasma current of between 10–15 mA is obtained, at which point the timer and/or thin film monitor should be turned on. The required thickness is discussed later in this chapter. Depending on the target material, the coating rate may be quite slow, and the metal atoms have very low impact energy on to the specimen. A slow sputtering rate favors small particle size. The temperature in the bell jar should only rise by about 10°C.

11.4.1.4 Final Stages of Sputter Coating

Once the coating is completed, the high voltage supply and vacuum pump should be turned off and the inert gas valve fully opened to allow the system to come up to atmospheric pressure. Avoid the temptation of allowing the chamber to come up to pressure using air. The coated specimens should be removed using the appropriate tools and gloved hands and stored in airtight boxes. Close the lid on the chamber and leave the system at atmospheric pressure. Finally, turn off the argon gas supply.

These procedures will ensure that samples receive an adequate coating. If the gas plasma is anything other than a light blue color it means that either the sample is out-gassing or that the argon is contaminated with oxygen, carbon dioxide, or water vapor. These contaminating gases will be broken down by the plasma to give reactive species that will damage the specimen. A good check on the effectiveness of a sputter coating procedure is to be able to either image cocoa butter crystals on a fractured M&M candy or the fine structure of dried milk powder in the SEM.

11.4.1.5 The Advantages of Plasma Magnetron Sputter Coating

The main advantage is that the highly controlled and reproducible procedure can be completed in about 10 min. The coating is multidirectional and the target alloys are sputtered congruently. As the metal atoms leave the target surface, they quickly collide either with the argon atoms or with each other. Their mean-free path of the sputtered particles is about 5 mm, and there are four to five collisions between leaving the target and arriving at the sample surface. This means that the metal atoms arrive at the sample surface from all directions, and with the exception of highly sculptured specimens, it is not necessary to tilt and rotate the specimen during coating. The congruent sputtering of alloys such as gold–palladium virtually eliminates epitaxial growth of the metals and ensures a small particle size in the coating layer. With modern plasma magnetron sputter coaters it is unusual to see etching and thermal damage on the sample surface.

11.4.2 Surface Coating by Ion Beam Sputtering

The use of focused ion beams (FIB) to investigate topography in the SEM was first described by Stewart (1962). Since then, FIB has been used to *erode* sample surfaces and the technology is important for nano-lithography in the fabrication of integrated circuits in the semiconductor industry. Although these large complex pieces of equipment could also be used for sputter *coating*, it would be rather like the metaphor of using a sledge hammer to crack a nut! Smaller pieces of equipment suitable for coating specimen for SEM are available and the Explorer ion beam coater shown in Figure 11.14 is from Denton Vacuum (www.dentonvacuum.com). Other supply companies include South Bay Technology Inc (www.southbaytech.com) and the ubiquitous ASaP Sample Preparation System from Fischione Scientific (www.fischione.com). The Fischione system includes

Figure 11.14. A turbo-molecular pumped Ion bean sputter coater for preparing thin layers of conductive films for scanning electron microscopy. Picture courtesy Denton Vacuum, Morrestown, New Jersey, USA

focused ion beam sputtering as part of an integrated specimen preparation designed for high resolution SEM and x-ray microanalysis. There are several recent publications on focused ion beam sputter coating and the recent books by Krishna (2001) and Giannuzzi and Stevie (2005) contain valuable information about the use of focused ion beams.

The target is eroded by a high energy argon plasma gas at low vacuum, while the sample is simultaneously coated in a high-vacuum environment separated by means of differential apertures. Figure 11.15 shows the main features of this type of coater. The focused high-energy beam of argon ions is set at an angle to the target that ensures the specimen is not exposed to high-energy electrons or ions that may cause surface damage. The inert gas plasma and target area are maintained at low vacuum 1 Pa (~10^{-2} Torr) in order to maximize target erosion rates and the sample is kept at high vacuum 10^{-4} Pa (~10^{-6} Torr) to minimize in-flight nucleation of the target atoms.

The positive ions and electrons generated in the gas plasma are deflected by large electrodes, which are placed immediately outside the low vacuum chamber. This process ensures that a

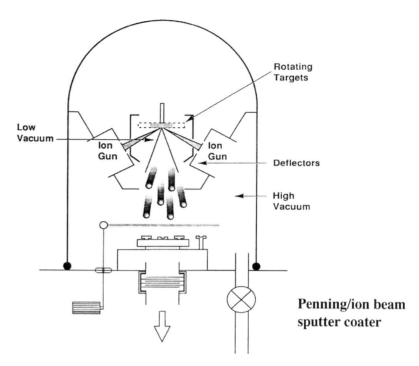

Figure 11.15. Diagram of a penning-ion beam sputter coater

virtually pure stream of metal atoms is directed onto the specimen. The mean free energy of the target atoms is an order of magnitude greater than that obtained by either thermal evaporation or plasma magnetron sputtering. This means that the metal atoms, which are not scattered by the residual argon atoms, are deposited onto the substrate with little lateral movement. These anchoring metal atoms act as nucleation centers from which the very thin films are formed with minimum vertical growth. Because the mean free path of the target atoms is so much longer than the target-substrate distance, the scattering advantage of the more conventional plasma magnetron sputter coater is lost and sculptured specimens must be rotated and tilted during coating.

These two ion beam sputtering processes cause minimal heating and etching and a wide range of targets, including carbon, may be used as source material. The ion beam and penning sputtering equipment is more complicated (and expensive) than the plasma magnetron equipment and their use is confined to very clean samples that have been prepared in a very clean vacuum system.

11.4.3 High-Resolution Coating Methods

A challenge to producing high spatial resolution images of non-conductive samples in the modern SEM is to be able to produce very thin surface conductive layers with very small grain size. Any film thick enough to guarantee nanoscale conductivity across a real surface will be too thick to permit high-resolution imaging. (D. Joy, 2008, personal communication).

It is not possible to resolve specimen details smaller than the grain size of the coating material. Both the evaporative and conventional plasma sputter coater can be used to prepare high-resolution images with barely detectable metal particles, provided correct target materials are used and the coating is carried out in a clean and orderly manner.

The basic aim for high-resolution coating is to use the thinnest possible continuous coating layer (more later) with the smallest grain size. The granularity of the film limits the image resolution, electrical conductivity, and mechanical properties of the thin layer. The coating layer should be thinner than the range of the electrons that carry the image signal, usually the SE1, most of which are generated in the first 1–2 nm of the surface. The film should form a continuous conductive surface, have very small grain size, and have sufficient density to provide image contrast. Thick films are unstable; they heat up and distort and obscure surface features.

Lowering the sample temperature ensures that much of the thermal energy of the arriving metal atoms is transferred immediately to the specimen, thus lowering the mobility of the arriving metal particles, and causes them to stay at their point of impact. In addition, the vacuum system should be made as clean as possible by using an oil-free pumping system, using strategically placed anti-contamination devices, and always backfilling the system to atmospheric pressure using a dry inert gas such as the boil-off from liquid nitrogen. The paper by Woodward and Zasadinski (1996) contains information that is useful when designing a high-resolution coating protocol.

11.4.4 High-Resolution Vacuum Evaporation Coating

These coating devices can be adapted to take advantage of the low grazing incidence and shadowing methods used with the TEM that employ target materials of platinum, tungsten, and a mixture of platinum–iridium–carbon to obtain 1-nm film layers. The technique devised by Hermann et al. (1988) involves moving an electron-gun–heated evaporation source of chromium or germanium back and forth through 90° during coating while maintaining the sample at 190 K and rotating it at 115 rpm. This process can produce continuous thin films of between 0.5 and 1.0 nm on *flat* surfaces. A simpler procedure is to immediately apply a very thin ~1 nm coating of carbon immediately after first coating the sample with a thin layer of metal. This reduces the re-crystallization of the thin metal layer. This procedure is also useful when using chromium and some of the refractory metals that oxidize very quickly; for example, 60% of a chromium film may quickly change to chromium oxide when exposed to air.

11.4.5 High Resolution Ion Beam and Penning Sputter Coating

For the very finest coating layers, ion beam and penning sputtering give the best results because they minimize the individual metal atoms from aggregating as they travel from their source to the specimen so they arrive at the surface as single (?) atoms. In contrast, as plasma-magnetron-sputtered and evaporated metal atoms travel from the target to the sample, they will aggregate and arrive at the surface as very small groups of atoms. For very high resolution images, this aggregation process may be sufficient to degrade resolution.

11.4.6 Osmium Metal Sputter Coating

When subliming osmium tetroxide vapor is introduced into the negative glow discharge of a plasma magnetron sputter coater, the molecules are broken down and the positively charged osmium atoms land on the specimen surface to give a very fine coating of osmium metal. Great care has to be taken when using this process because of the toxicity of osmium tetroxide.

A specially designed osmium coater, the OPC-80T shown in Figure 11.16, is available from Structure Probe Inc (www.2spi.com). Although it is recommended that the osmium coater is used either in a fume cupboard or vented directly to the outside, there are

Figure 11.16. Osmium Plasma Coater Model OPC-80T. Picture courtesy Structural Probe Inc., West Chester, Pennsylvania, USA

sufficient safe guards built into the equipment to allow it to be used at a laboratory bench. This hermetically sealed equipment is fitted with a series of automatic interlocks that go through the following sequence once the sample is loaded into the coater:

1. The argon gas supply and the line to the vacuum pump are checked in the ways described earlier in this chapter.
2. The specimens are loaded onto the sample holder.
3. A glass ampoule containing 0.1 g osmium tetroxide crystals is placed in the internal osmium container, which is then sealed. This size ampoule will give enough osmium for 10–15 runs.
4. The bell jar is pumped out and flushed with argon several times.
5. The bell jar is evacuated to a pressure of 0.5 Pa (5×10^{-1} Torr).
6. An automatic sequence is then initiated, which:
    a. If necessary, breaks open a new glass ampoule of osmium tetroxide crystals.
    b. The osmium tetroxide internal container is opened.
    c. A pre-set amount of osmium tetroxide vapor is introduced into the bell jar.
    d. The high power is turned on to a pre-set value. The sample is sputter coated with osmium metal for a pre-set time, which gives a metal layer of 1–2 nm thick, which is enough to form a continuous conductive layer.
    e. The osmium tetroxide internal container is re-sealed.
    f. The high power is turned off.
    g. The bell jar is thoroughly flushed with argon gas, which is pumped out via a trap, which absorbs any remaining osmium tetroxide vapor.
7. A light comes on once the automatic sequence is finished.
8. The coated samples may now be removed and any remaining osmium tetroxide crystals or vapor remain safely sealed in the internal container ready for the next coating sequence.

The results are impressive and Figure 11.17 shows a series of glass slides each coated with the minimum amount of material to make them conductive. The gold, gold–palladium, chromium, and platinum were deposited with a plasma magnetron coater; the carbon was vacuum evaporated and the osmium deposited with the osmium sputter coater described in the preceding.

The osmium metal coating layer has a very small grain size and even at a magnification of $\times 10^6$ in the TEM, no coating details can be seen on an image of thin holey Formvar coated support grid. (P. Echlin, 2004). Although osmium is not as highly conductive as the noble metals, it is the element with the highest density. A study by Osawa et al. (1999) demonstrated that osmium metal

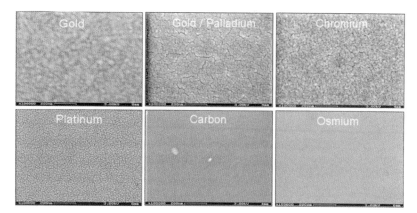

Figure 11.17. A comparison of the grain size of six different coating materials. The minimum amount of material was deposited to give a continuous layer on a clean glass substrate. The gold, gold-palladium, chromium and platinum were applied using a plasma magnetron sputter coater. The carbon was applied using a vacuum coater and the osmium was applied using an osmium plasma sputter coater. The SEM images were recorded at 3.0 keV, and ×100,000 magnification. The picture width is 1 μm. Picture courtesy Jim Young, Pacific Northwest National Laboratory, Richland, Washington, USA

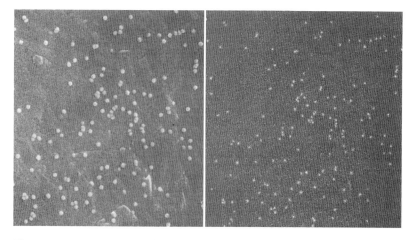

Figure 11.18. 15 nm gold-conjugated goat anti-mouse IgG antibody air dried on a polished carbon surface and coated with approximately 10 nm of osmium. The SE (left) and BSE (right) are images of the same specimen. Magnification ×50 K. Picture courtesy Osawa et al. (1999)

thin coating may be used in conjunction with backscattered electron imaging of colloidal-gold-labeled antibodies. Figure 11.18, from this study, shows SE and BSE image of a sample coated with osmium metal.

The osmium plasma coater is designed for preparing specimens for very high spatial resolution in an FESEM. It is not recommended for routine SEM. Figures 11.19 and 11.20 show SE

Figure 11.19. SE image of a hole in a photo-resist layer on silicon. Image on the left is coated with 5 nm osmium, the right hand image is uncoated. Images at ×50 K and 1 keV. Picture courtesy Structural Probe Inc., West Chester, Pennsylvania, USA

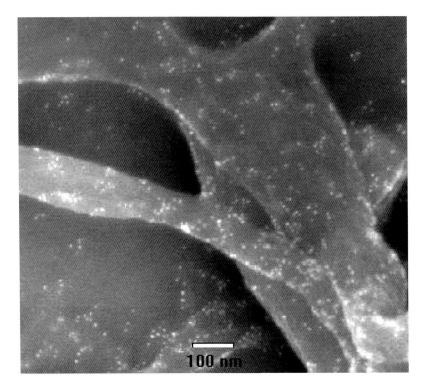

Figure 11.20. BSE image of human astrocoma tissue, double labelled with 5 and 10 nm gold probes. Coated with 2 nm of osmium. Picture courtesy Structural Probe Inc., West Chester, Pennsylvania. USA

and BSE images of two different types of material coated with a thin layer of osmium metal.

11.5 Coating for Analytical Studies

The equipment used in coating for analytical studies is no different from that used for structural studies. The only differences center on the choice of the target material. Ideally, the coating layer

Table 11.9. Physical properties of some elements which may be used to coat non-conductive sample for x-ray microanalysis in the SEM

Property	Beryllium	Carbon	Aluminium	Chromium
Density (Kgm^3)	1800	2300	2700	7200
T-Conduction (W.cm)	2.00	1.29	2.37	0.93
Resistivity (µohm.cm)	4.27	3500	2.83	13.0
Evaporation-T (°C)	1247	2727	1002	1177
X-ray energy (eV)-Ka	109	227	1487	5415
-La				573

should not contribute to the x-ray spectrum, must be sufficiently thin not to mask or substantially overlap the emitted x-ray photons or attenuate the incoming electrons. For these reasons, carbon is used as a coating material in spite of its poorer electrical conductivity and high heat of evaporation. This presents no problem with inorganic and tough organic samples, but is not the best approach for many organic and all biological material.

Beryllium, aluminum, and chromium may be used as coating metals, and beryllium is probably the best material. It has good thermal and electrical properties, negligible x-ray interference and makes little contribution to the x-ray background. Unfortunately, it is very toxic when in the finely particulate form. Marshall and colleagues made extensive use of beryllium as a coating material in connection with their analytical work on insect tissue. Any one considering using beryllium as a coating material in connection with x-ray microanalysis should first consult the papers by Marshall and Carde (1984) and Marshall et al. (1985) on details of the very strict procedures and precautions that must be used with this metal. Some of the properties of these three metals together with the properties of carbon are shown in Table 11.9.

Provided very thin continuous coating layers of conductive metals are formed, it is possible for use such materials whose x-ray spectra overlap the x-ray spectra of the elements being analyzed in the sample. The following two examples show how this may be achieved.

11.5.1 Gold Evaporative Coating on a Polished Mineral

The M x-ray lines of gold (2,196–2,313 eV) overlap the K lines of Si (1,840); P (2,144) and S (2,470), which might be found in minerals and, at a first glance, gold should not be used as a coating material. However, there is a case for using a very thin layer (0.5–1 nm) of gold on samples that are analyzed at high voltages (10–20 keV). The incoming electrons are not attenuated, the spurious emitted gold x-rays are small but quantifiable, and the software on modern x-ray microanalysis equipment allows the gold x-rays to be removed from the total spectrum. The advantage is that gold is 1,500 times more electrically conductive than carbon.

11.5.2 Chromium Sputter Coating of Fresh, Frozen Hydrated Tea Leaves

Studies by the author were carried out on the location and concentration of the elements Mg, Si, and Al in fresh green tea leaves. Small samples of fresh leaves were quench frozen in melting nitrogen and immediately inserted into a high vacuum, low temperature device attached to the side of the SEM. The frozen samples are fractured and coated with a 1–2 nm layer of chromium using a plasma magnetron coated and very high purity argon. The samples, although under high vacuum, were passed via a gate valve to the pre-cooled cold stage on the SEM and subject to x-ray microanalysis at 3–5 keV and a beam current of no more than 200 pA.

The x-ray K line of chromium is 2,622 eV, which did not effect the K lines of Mg (1,254 eV), Al (1,487 eV), and Si (1,740 eV). The principal element in the frozen material is the oxygen of the frozen water that makes up 90% of the tea leaf. The K line of oxygen is 525 eV, which is perilously close to the L line of chromium at 573 eV and formed a slight shoulder on the large oxygen x-ray peak. It was possible to use spectral devolution techniques to remove the contribution of the chromium x-rays from the huge oxygen peak. In any event the small chromium addition to the oxygen peak did not really matter because oxygen concentration was not part of the quantitation but served to show that the sample was fully hydrated during the x-ray microanalysis. The advantage is that chromium is 270 times more conductive than carbon and can be sputter coated at 130 K.

11.6 Coating for Low Temperature Microscopy and Analysis

This subject is considered in some detail in the book by Echlin (1992), and additional details may be found in the papers by Chen et al. (1995) and Walter and Muller (1999). The coating procedures are the same as those described elsewhere in this chapter, but with the added proviso that the sample must remain at or below the glass transition temperature of the specimen. For example, the T_g of water is 130 K. This means that the coating procedure(s) must be carried out in a completely anhydric environment and the sample is not subjected to any thermal stress. For example, carbon evaporation causes surface relegation on frozen-hydrated specimens. The best results are obtained using a specially designed plasma magnetron sputter coater that can be first evacuated to a high vacuum using a turbo-molecular pump and then very high purity argon gas to form the plasma. Gold and platinum are the best target for SEM images and chromium for x-ray microanalysis. Carbon should not be used because it is essentially an insulator at the low temperatures used for this type of microscopy and there is the ever-present problem of the high temperature needed for carbon evaporation.

11.7 A Guide to Coating Thickness for SEM and X-Ray Microanalysis

The single rule regarding coating thickness is to apply the minimum thickness in order to obtain the maximum information from the specimen. The corollary is that if a noble metal coated sample shines like gold, use it for self-adorning jewelry, not for the SEM. The sequence of images in Figure 11.21 shows that if the coating layer is too thick, it will obscure the surface detail. If the coating layer is too thin and/or discontinuous, the surface will charge and the sample will be damaged. The conductivity of a very thin metal film is less than that of the bulk metal, although it is assumed that once a continuous layer of material forms on the specimen, there is a significant increase in its conductivity.

Various attempts have been made to try and establish the threshold film thickness of a continuous film for different coating materials. Although the data is very variable and depends on the sample and exactly how the thin film is deposited, the following numbers emerge. Tungsten 1–2 nm, osmium 1–2 nm; chromium 1–2 nm; platinum 1–2 nm, silver 3–5 nm, gold–palladium 3–4 nm, gold 6 nm, and carbon 10 nm. Paulson and Pierce (1972) have shown that discontinuous films examined at 20 keV and very low beam current, can conduct a limited current by electron tunneling between islands of evaporated gold with an estimated thickness of 2.5 nm.

11.7.1 Film Thickness Calculated Before Coating

11.7.1.1 Evaporative Coating

The thickness of a layer deposited on a specimen surface by thermal evaporation, can be estimated from knowing the mass of the starting material that will be totally evaporated. In high vacuum thermal evaporation, one can assume that all the vaporized metal molecules leaves every part of the material being in a straight line, without a preferred direction and pass to the specimen surface without colliding with residual gas molecules. By allowing for the vapor incident angle at the specimen, and assuming that all the incident metal vapor molecules have the same condensation coefficient, the *average* coating thickness on the sample may be calculated. Equation 11.2 can be applied to a flat, no rotating surface oriented at an angle q to the source direction, and used

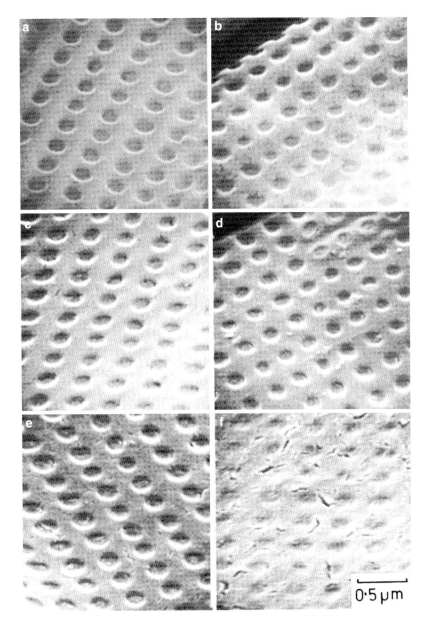

Figure 11.21. A series of SE images of cleaned diatom silica frustules Evaporative coated with an increasing mass thickness of gold. a) 2.5, b) 5.0, c) 10, d) 20, e) 50, and f) 100 nm. The optimal coating thickness is between 2.5 and 5 nm after which there is a progressive decrease in image quality. Images taken at 10 keV

to calculate the coating thickness from the mass of the material evaporated.

$$T = \frac{M}{4\pi R^2 p} \cos\theta \qquad (11.2)$$

where T is the coating thickness in centimeters, M is the mass of evaporant in grams, p is the density of the evaporated material (g/cm^3), R the distance from the evaporated source to the sample in centimeter. and q is the angle between the normal to the specimen and the evaporation direction.

With this equation, the thickness can be calculated to ±50% if the geometry of the evaporation rig and substrate is known and if the material in question evaporates slowly and does not alloy with the heater. It is assumed that evaporation is from a point source and occurs uniformly in all directions. An accurate measurement of thickness is further complicated by the uncertainty in the sticking coefficient of the evaporated material on the substrate and the deviation of the film density from the bulk density. This sticking coefficient is rarely better than 75% and any thickness calculation using the formula above must be multiplied by 0.75.

As an example, consider a carbon evaporative coater that operates by completely evaporating a 1 × 3 mm carbon rod which weighs 4.7 mg. If the specimen is 10 cm from the carbon source and perpendicular to the distance of evaporation, the thickness of carbon on a flat glass cover slip would be 7–21 nm. For modern SEM and x-ray microanalysis this degree of uncertainty is intolerable.

11.7.1.2 Sputter Coating

It is possible to calculated the film thickness T deposited by a plasma magnetron sputter coater using Equation 11.3:

$$T = \frac{C.V.t.K}{10} \quad (11.3)$$

where C = the plasma discharge (mA), V = keV, t = time in minutes, and K is a constant depending on the bombarding gas and the target material. This constant must be calculated for each specimen coater, target material, and the bombarding gas, usually argon. As an example, a gold:palladium, 4:1 target, plasma magnetron sputtered with argon at 10 mA and 1.8 keV for 4 min would give a thickness on a flat glass cover slip of 5–10 nm. Although this calculation is a little closer than that in Equation 11.3, it is still not sufficiently precise for modern SEM. It is quite clear that calculating the film thickness before coating is both cumbersome and inaccurate and the estimates for film thickness are now made during the actual coating process.

11.7.2 Film Thickness Calculated During Coating

It is possible to calculate the mass density of the coating layer by measuring the ionization that occurs when vapor molecules collide with electrons or by measuring the force impinging parti-

Figure 11.22. The type of quartz thin film monitor used to measure thin film thickness during coating of SEM specimens. The monitor is at the centre with an area of 5 mm² and the connections lead to an exterior monitor

cles exert on a surface. Mass-sensing devices may be used for all evaporated or sputtered materials by determining the weight of a deposit on a microbalance or by detecting the change in oscillating frequency of a small quartz crystal on which the material is deposited. Figure 11.22 shows the type of small quartz thin film monitor that can be placed in a thin film coating device.

If needed, the details of all these methods can be found in the books by Maisell and Glang (1970) and Smith (1995). For SEM and x-ray microanalysis, the most practical approach is to use a quartz thin film monitor.

11.7.2.1 Quartz Thin-Film Monitor

A quartz thin film is made to resonate and its frequency is directly related to the mass of material on its surface. The frequency of the crystal is measured before and after coating, and the decrease in frequency gives a measure of the film mass thickness. It is important to remember that these monitors do not produce an absolute measurement, because they average out discontinuous thin metal coats. The monitor measures the *average mass-thickness*; that is, if the film is thin and/or discontinuous, is much less than the *average metric thickness*. As the film thickness increases, the two forms of measurement come closer to giving the same value.

A typical sensitivity value for a crystal monitor is a 1-Hz frequency change for an equivalent film of 0.1 nm carbon, 0.13 nm aluminum and 0.90 nm of gold. The uncertainty of the devices made specifically for evaporative and plasma magnetron sputter coaters is of the order of ±0.1 µg cm^{-2}. The thin film monitor is placed as close as possible to the sample and all that is necessary is to program the device with the density of the metal being used as the coating material. Accurate measurement of thin-film thickness can only be made using thin quartz thin-film monitors in which the quartz disk is cut relative to the main axis of the crystal and that the monitor is kept at a precise temperature. These devices are quite expensive, and as shown later, are not necessary for measuring thin-film thickness for SEM and x-ray microanalysis.

It is naïve to equate a thin-film measurement given on a planar quartz crystal with that on a complexly sculptured specimen being rotated and rocked during the coating process. Various studies—see for example Peters (1980, 1986), Clay and Peace (1981) and Hermann et al. (1988)—have shown that there is a wide discrepancy between the value given by a quartz thin-film monitor and the actual amount of material that lands on the specimen's surface. What do these discrepancies mean to the microscopist concerned with providing a reasonable coating thickness to ensure maximum specimen-image information transfer?

For thick continuous films (i.e., more than 15 nm), the average thickness given by a quartz thin-film monitor is fairly close to the average metric thickness that may be measured on a planar surface. If accurate measurement are required on films less than 10 nm in thickness, the indicated average mass–thickness given by the thin-film monitor should be checked by some of the independent means such as light absorption, transmittance, reflection, and interference by various optical monitors or by means of situ resistance measurements or capacitance. The papers by Flood (1980) and Johansen and Namork (1984) describe these different techniques.

11.7.2.2 Practical Use of a Quartz Thin Film Monitor

The most sensible way to use the monitor is to set up a standard coating procedure for a given group of samples. Have a set target-substrate distance, residual gas pressure, target material, source and substrate temperature, keV and current, etc. The single variable is time. Using the Goldilocks approach, a set of similar samples are then coated for different time periods and the thickness monitor reading is recorded. The specimens are then examined in the SEM at a set KeV and beam current. The most informative image indicates the best way to coat the sample. If the coating is too thin, the samples will show incipient charging, if it is too thick, fine surface details will be obscured. For example, P. Echlin (2004) in a continuing study on drug particles in a FEI FESEM used the following standard conditions:

Plasma magnetron sputter coater. High purity argon gas, gold:palladium target, 65 mm between target and specimen, 1,800 eV, and 10 mA beam current and allowed the thin film monitor to record a thickness of 4–5 nm.

Scanning electron microscope. Operated at 3 keV, 10 pA beam current, working distance 15 mm and images recorded from ×100 to ×50,000.

The images showed no charging, good contrast and provided the necessary details required for the work. The actual thickness of the thin gold:palladium film *on the sample* is unknown.

11.7.3 Film Thickness Calculated After Coating

The most accurate measurements of film thickness may be made on the films themselves. These methods are based on optical techniques, gravimetric measurements, and x-ray absorption and emission. Multiple beam interferometry is the most precise and, depending on the method used, can be as accurate to within 1–2 nm. The Fizeau method, which can be used to check film thickness, involves placing a reflective coating on top of the deposited film step and measuring a series of interference fringes. The film thickness can also be measured by sectioning flat pieces of polymerized resin on which a coating layer has been deposited and measuring the metal thickness in the TEM. The accuracy of this method depends on being able to section the resin and record the image a right angles to the metal deposit.

One of the easiest and, surprisingly, sensitive methods is to examine the surface of either a half-covered 10 mm^2 piece of brilliant white paper or a glass cover slip, placed close to the sample during coating. Depending on the coating thickness, the piece of paper will show various shades of gray and a distinct reflective and/or transmitted color on the glass cover. For most specimens, a light gray tone on the white paper indicates a satisfactory coating thickness for the noble metals, aluminum, chromium, and carbon. On a glass cover slip, a carbon layer visible as a chocolate color and a reddish-brown gold layer by reflected light and blue-green by transmitted light are satisfactory. Table 11.10, shows that the varying thicknesses of carbon films can be judged by the changes in the reflected color on a flat, highly polished brass surface For aluminum coating, sufficient metal is deposited when the layer is a deep blue to transmitted light.

The true thickness could be measured using the Fizeau technique discussed previously. For SEM and x-ray microanalysis the actual film thickness is less important than the thickness of the coating that provides the optimal information about the sample. For very high resolution SEM, Wepf et al. (1991) used wavelength-dispersive x-ray microanalysis to measure the composition and hence the average mass thickness of Pt–Ir–C films.

Table 11.10. Estimation of carbon film thickness from the colour of an evaporated layer deposited on a highly polished brass surface

Film Colour	Thickness
Chocolate	10 nm
Orange	15 nm
Indigo	20 nm
Blue	25 nm
Blue-Green	30 nm
Green-Blue	35 nm
Pale Green	40 nm
Silver-Gold	45 nm

11.7.4 General Guideline to Specimen Coating

These guidelines are bases on the following premises:

1. Modern SEM and x-ray microanalysis is carried out using field emission microscopes.
2. Most SEM is carried out at low voltages ~5 keV and low beam current ~10–25 pA.
3. A substantial amount of SEM is carried out at magnifications between ×1 and 25 K.
4. Most x-ray microanalysis is carried out at between 5–20 keV and 200 pA to 200 nA.
5. Plasma magnetron sputter coating and derivatives, are used more than evaporative coating.

11.7.4.1 For Very High Resolution Images

A 0.5–0.8 nm film is recommended using a penning/ion beam sputter coater with targets of platinum, tantalum, iridium, tungsten, or chromium. All internal parts of the coating equipment must be scrupulously clean, use stainless steel gas and vacuum lines with metal–metal seals and very high purity argon. Alternatively, one can use electron beam evaporation to apply the same thickness of a platinum–iridium–carbon mixture. The question remains whether these very thin films of material are continuous or discontinuous.

11.7.4.2 For High Resolution Images Above ×50 k

A 1- to 2-nm film of either a platinum or chromium layer using a plasma magnetron sputter coater with high purity argon. There is an increasing amount of evidence that suggests that a 1-nm osmium metal sputter coated film could be used to prepare high resolution images.

11.7.4.3 For Medium Resolution Images at Between ×5–×50 K

A 3- to 4-nm layer of gold–palladium using a plasma magnetron sputter coater with high purity argon.

11.7.4.4 For Low Resolution Images Below ×5 K

First try imaging the sample at low voltage and beam current without coating. If necessary, deposit a 5- to 6-nm layer of gold–palladium using a plasma magnetron sputter coater with high purity argon. These thicker layers of coating may well break the Gold Adornment Rule but this is unlikely to degrade the quality of the low magnification images.

11.7.4.5 For X-Ray Microanalysis

A 10-nm layer of evaporated carbon or 3–4 nm of a plasma magnetron sputter coated conductive metal compatible with the elements being analyzed.

These are only guidelines and it is import to experiment with materials, film thickness, and mode of applying the thin film. Always start with thin conducting layers before proceeding to thicker deposits. Highly sculptured specimens need more coating material than flat samples and specimens examined at high voltage need a thicker coating layer than those examined at low voltage.

11.7.5 Removing Coating Material from Specimens

Having taken all the trouble to deposit a coating layer on a sample, it is occasionally necessary to remove it. This may be necessary with unique specimens such as paintings, sculptures, pottery, jewelry, ceramics and furniture; unique archaeological findings, fossils, and forensic specimens. Jewelers, museum staff, art dealers, antique owners, and law enforcement officers are sometimes proprietary about their samples and become anxious about the unique properties of their specimens. The first approach is not to coat the specimens and to examine them at low voltage and low beam current.

If it has been necessary to cut, lap, polish, and then coat a unique hard sample prior to microscopy and analysis it can never be returned to its original pristine state, although the coating can be easily removed by returning to one of the final polishing operations. If the surface of the original sample was rough and thinly coated, the only way to remove the coating material is by the judicious application of plasma or ion beam etching and, as the last resort, use chemical means. These approaches are not to be taken lightly.

Where it is known that the precious, unique, nonconductive sample has to be returned unadorned, then it is best to use aluminum, gold, gold–palladium, or carbon as the coating material as they are somewhat easier to remove than other coating materials. Sylvester-Bradley (1969) removed thin layers of aluminum from geological samples by immersing the sample surface in a freshly prepared solution of dilute alkali for a few minutes. Sela and Boyde (1977) used a cyanide solution to remove gold films and Crissman and McCann (1979) used a 10% solution of $FeCl_3$ in ethanol to remove gold–palladium. It should be noted that these and other chemical methods should only be used as a last resort because of the co-lateral damage they may cause to the sample.

Finally, devices such as the Model 1030 Automated Sample Preparation System marketed by Fischione Instruments (www.fischione.com) may be used equally well to *remove* selected material from sample surfaces as to *deposit* selected materials on the surface.

Carbon layers can be removed by reversing the polarity in the sputter coater to allow the argon ions to strike the specimen. Care must be taken not to damage the sample.

11.7.6 Some Visible and Measurable Indication of Inadequate Specimen Coating

Although the general subject of artifacts and damage is discussed in the next chapter, it is more useful here to briefly discuss the causes that led to a failure of surface charge elimination. If the coating processes has been carried out properly, artifacts and damage are rarely seen in the images and in the analytical data.

11.7.7 Consequences of Incomplete Evaporative Coating

11.7.7.1 Thermal Radiation

The intensity of thermal radiation reaching the specimen depends on the source of heat and the source-to-specimen distance. The heat of radiation can be diminished by using a small source and/or moving the specimen further away. The best practical solution is to use a small source at high temperature, have at least 150 mm between the source and the sample and employ a mechanical shutter to control the process. Provided the specimen is adequately shielded from the target and the shutter opened only at the working evaporation temperature, little damage will occur. An additional caution is to have a cooled shutter that rotates at about 100 times a second during the actual source evaporation.

Thermal artifacts appear as minute holes, surface distortions, and smooth, micro-melted areas on inclined fracture faces on biological and organic material that are bombarded vertically.

11.7.7.2 Contamination

This is due, primarily, to dirt and volatile substances in the vacuum system being deposited on the specimen and it is for this reason that care must be taken to clean the system before use. For very-high resolution studies, the most effective way to reduce this problem is to surround the specimen with a cold surface such as a liquid nitrogen trap. Contamination may be recognized as uneven coating and hence charging, as small randomly arranged particulate matter and in the most extreme situations, as irregular dark areas on the specimen image.

Agglomeration effects: Agglomeration of the evaporated material occurs to some extent with most metals and is a result of uneven deposition of the evaporated compound. The cohesive force between the individual evaporated molecules is greater than the forces between the film molecules and the sample surface. Because of geometric effects, rough surfaces are particularly difficult to coat evenly, and it is inevitable that parts of the specimen which protrude receive more coating that crevices and holes. This particular problem can be minimized by rotation and tilting the sample during the coating process.

11.7.7.3 Film Adhesion

Poor film adhesion is associated with hydrocarbon and water contamination and, in the case of some plastics, with the presence of a thin film of extruded plasticizer. Discontinuous and poorly adhesive films are recognized by a "crazed" surface appearance and have a tendency to flake easily. In the microscope, the image brightness and charging occur on isolated "islands" of the sample not in contact with the rest of the film.

11.7.8 Artifacts and Damage During Sputter Coating

There is no doubt that the early sputter coaters damaged the surface of delicate samples. These problems are only rarely seen with the modern recent plasma magnetron coaters, particularly if they are connected to a turbo-molecular pump. However, some artifacts and damage may still occur.

11.7.8.1 Thermal Damage

The target material heats up during the sputtering process. This heating is minimal when using the noble metals but can get quite high with other less conductive metals such as tungsten and tantalum. The only way to control the thermal input into sensitive sample such as biological and organic materials is as follows:

1. Strike a balance between slow coating speeds, which favor smaller film particles, and the increased time this exposes the sample to thermal input.

2. Rotate and tilt the sample during coating.
3. Use targets with high sputtering speeds, i.e., gold sputters five times faster than tungsten (see Table 11.6). Thermal damage is manifest as melting, pitting, and in extreme cases, sample destruction.

11.7.8.2 Decoration Artifacts

Leaving aside the agglomeration effects discussed earlier, decoration artifacts can be a serious problem in some high resolution images. Peters (1985) has made an extensive study of this problem and has shown that decoration artifacts are a feature of high surface-mobility elements such as the noble metals, which form a continuous layer only at a thickness of 3–6 nm. Even if thinner, discontinuous films are used, both gold and platinum particles move from the sides of surfaces and aggregates on top of the surface molecules. It is possible to see this phenomenon in images of thin projecting specimens. There is a thicker coating layers at the very tip compared with the sides. This is an artifact not a genuine structural feature. This phenomenon does not occur with chromium, which has a much lower surface mobility.

11.7.8.3 Surface Etching

This is a potential problem when stray bombarding gas ions or metal particles hit the sample surface with sufficient force to erode it away. This erosion appears at high resolution as small pits on the sample surface.

11.7.8.4 Film Adhesion

This is less of a problem with sputter coated films than with evaporated films and is probably due to the fact that the metal particles penetrate the sample surface. However, sputter-coated samples and, indeed evaporated coated samples, should not be exposed to wide excursions of temperature and humidity after coating and before imaging and analysis, which can cause the samples to expand and contract with a consequent rupture of the surface film.

11.7.8.5 Contamination

Just because sputter-coating is carried out at a much lower vacuum pressure than evaporative coating does not mean one can afford to be cavalier with cleanliness. The same rules that apply to evaporative coating apply to sputter-coating.

11.8 Suggested Procedures for the Six Different Types of SEM Specimens

Unlike the finale of the previous chapters, there are no specific suggestions for each of the six different types of specimens, with three notable exceptions. If the sample surface is not sufficiently conductive to allow the incoming electron beam to easily flow to earth, it will need to be made conductive by one or more of the different ways considered in this chapter. The two exceptions are as follows.

11.8.1 Metals, Alloys, Metalloids, and Metallic Materials

These materials are good conductives and should not require any need for charge elimination. However, the rough surfaces of some alloys, metalloids, and metallic materials may present a problem, particularly if these materials are a discrete but integral part of a non-conductive material. It is best to first try to image such materials without using any of the charge elimination procedures but, as a last resort, a very thin layer of one of the noble metals may be applied. Similarly, metal particles in a ceramic matrix may need coating with a thin conductive film.

11.8.2 Hard Dry Inorganic and Organic Samples, Polymers, and Biological and Wet Samples

With one exception, carbon should not be used as a coating material for any of these types of samples. The only exception is to use it for the x-ray microanalysis of hard dry inorganic materials such as rocks, minerals, etc. Carbon should not be used for SEM imaging.

11.8.3 Microelectronics, Packages, and Semiconductors

These types of samples which contain a mixture of conductive and non-conductive metals, ceramics, and polymers, present a challenging group of problems for SEM and x-ray microanalysis. They consist of integrated circuits made up of metallized and passivated regions, electrical attachments to the integrated circuits, and their attachment within the package and to the printed wiring board. Unlike most other specimens examined in the SEM much of the interest centers on observing the dynamic changes in their structure in both the passive and active state. An additional thin coating layer of metal could create all sorts of problems and a number of new approaches have been developed.

It is now possible to apply continuous thin metal films (2 nm or less) of iridium, tungsten, or platinum, which provide a perfect ground plane suitable for very high resolution SE1 imaging. Such thin films can also be used in association with low voltage x-ray microanalysis because the metal layers are too thin to give any appreciable contribution to the spectrum and charging effects are control.

The recent book edited by Zhou and Wang (2006) contains a wealth of information about preparing nanotechnology specimens for SEM and x-ray microanalysis.

1. Secondary electron images of the specimens may be examined uncoated in the SEM using low pressure 10–50 Pa (76–380 × 10^{-3} Torr), at a sufficiently low voltage at which a balance between positive and negative charging can be found. Backscattered imaging also may be used.
2. The SEM can utilize the voltage contrast, electron beam induced current, and electron backscattered diffraction modes to provide important information about these types of samples.
3. Dual-beam microscopes which have electron and ion beam columns on the same instrument provide the best of both worlds. The highly controlled ion beam may be used to precisely cut or cross-section specimens, deposit metal layers at precise locations on parts of the sample and provide spatial resolution as good as 5-nm. The secondary electrons can provide very high resolutions images. From the charging point of view it is possible to inject several different gases at the ion beam impact point to deposit both conductive and insulating materials, or selectively etch the cross-section of the sample surface. The dual beam microscope is the ultimate mini-laboratory. Further details of these instrumental methods are given in the paper by Prewett (1992) and in Chapter 11 in the book by Goldstein et al. (2004).

Dual beam microscopes are very expensive, but there is one other less expensive way to overcome the charging problem in microelectronic devices. Instruments are now available that combine the benefits of plasma cleaning and ion beam technology to prepare specimens before they are examined in the SEM. One of these instruments is the Model 1030 Automated Sample Preparation System marketed by Fischione Instruments (www.fischione.com). This piece of equipment has already been mentioned several times elsewhere in this book.

The system can take a 25-mm diameter, 25-mm deep sample, which has been prepared by cleaving, grinding, cutting and/or sectioning, through plasma cleaning, ion beam etching, reactive ion beam etching, reactive ion etching, and ion beam sputtering. The system is fully programmed and operates in an oil-free vacuum provided by a turbo molecular drag pump backed by a multistage diaphragm pump. Up to four different targets from W, Cr, Pt, Ir, Ta, and C can be used and up to three different process gases can be used to etch the sample surface. The system has been specifically designed to prepare specimens for advanced high resolution field emission SEM.

12

Sample Artifacts and Damage

The sole reason for using a scanning electron microscope is to obtain accurate, precise and reproducible information about their structure and chemical identity. We seek information either to confirm and extend our existing knowledge about an object *or* investigate a new and unknown object. The information we obtain is either in the form of a picture (image) or as files of numerical data. We need to be able to validate this information because the processes of obtaining images and data using the SEM are usually very invasive and totally alien to the environment in which we and our specimens exist. We must be satisfied that the procedures used to obtain information do not damage the object or introduce artifacts.

Damage is an unexpected and irreversible change in the object and can occur before and during microscopy. In many cases, damage is very obvious in an image and some examples are shown later in this chapter. However, sometimes the damage is less immediately obvious.

Artifacts are perceived structural distortions or misrepresentative chemical changes to the original object that arise as a consequence of the techniques used in preparing objects for subsequent microscopy and analysis. Artifacts are frequently not immediately obvious.

A good example of an artifact is seen in the images of biological membranes. The "best" high resolution images were first obtained by using specimens that were first chemically stabilized, infiltrated with anti-freeze agents, and then quench cooled. The membranes were exposed by a freeze-fracture process, freeze-dried, and coated with a very thin layer of a noble metal. A large amount of data was obtained that contributed immeasurably to our understanding of the structure and function of membranes. The preparative methods were continually refined and it was found that the antifreeze agents that effectively prevented ice crystal damage introduced image artifacts.

Membrane studies now rarely use antifreeze agent and rely more on high pressure cooling to obtain vitreous samples.

Thus, it is imperative that we know exactly what we have done to an object in order to obtain valid information about the sample. Although we use the rigor of statistical analysis to validate the veracity of numerical data, this alone is not enough to validate this information.

There is no single indication in an image that would suggest there is either damage or artifact but one should be very suspicious of an image that is composed solely of absolute black and white areas. Each of the six groups of samples considered throughout this book has their own catalog of faulty images and spurious data. Metals and hard inorganic materials are less susceptible to damage and artifact than dry organic material, polymeric, biological samples, and wet materials.

Table 12.1 provides a list of some of the image aberrations and artifacts that can arise as a consequence of faulty procedures that occur during the early phases of sample preparation of polymers, biological, and wet samples.

Table 12.1. Faulty secondary electron images of polymers, biological and wet samples due to artifacts and damage arising from incorrect sample preparation

SEM image aberration or damage	Probable cause
Bubbles and blebs on the surface	Incorrect washing and hypertonic stabilization
Burst bubbles	Incorrect stabilization fluids
Cell surface depression	Incipient air drying
Crystals on the surface	Insufficient removal of stabilizing electrolytes
Cell detached from the surface	Poor rapid cooling, poor dehydration
Cell contents clumped and coagulated	Poor stabilization and dehydration
Deformation of otherwise thin layers	Faulty cutting or fracturing
Directional surface ridges	Sectioning or planning damage
Disruption of whole cells and tissue	Poor stabilization, freezing, or dehydration
Disruption of cell surface	Poor washing and stabilization
Disruption of cell shape	Air drying dehydration
General sample shrinkage	Hypotonic stabilization and washes
Gross damage to sample	Poor stabilization, microbial decay
Large surface particles	Disintegration of specimen dirt, and dust
Microcrystalline deposits	Faulty buffer, stabilization, or stain
Regular splits and cuts	Thermal contraction in the coated sample
Ruptured cell contents	Poor critical point drying
Separation of cells in a tissue	Incorrect stabilization solutions
Smooth surface on a featureless sample	Incorrect dehydration
Stretched cell surfaces	Hypotonic stabilization solutions
Strands and sheets of surface material	Incorrect washing
Surface folds and ridges	Incorrect pH of stabilization solution
Surface sloughing	Dehydration from organic liquids
Surface with unexpected irregularities	Incorrect time for sample stabilization
Swelling of the sample	Hypertonic buffers and stabilization solutions

Table 12.2. Image artifacts and spurious x-ray date due to specimen surface charging, contamination or beam damage

Image quality and/or x-ray spectrum	Probable cause
Background distortion	Surface charging
Bright areas on surface	Surface charging
Dark areas on surface	Surface charging
Bright areas in cavities	Surface charging
Bright banding on surface	Surface charging
Whole image distorted	Surface charging
Whole image shift	Surface charging
Image movement	Surface charging
Bright raster lines	Surface charging
Dark halo around bright sample	Surface charging
Growing bubbles on the surface	Beam damage
Cracks on the surface	Beam damage
Small holes on the surface.	Beam damage
Surface shriveling	Beam damage
Surface movement	Beam damage
Growing holes on the surface	Beam damage
Particles on sample surface	Too much coating
Increased particle dimensions	Too much coating.
Obscured surface detail	Too much coating
Large surface cracks	Too much coating
Speckled surface	Contaminated coater
Surface crazing and etching	Incorrect coating
Surface melting	Thermal damage while coating
Raster square remains of surface	Surface contamination? or Surface charging?
Dark and bright raster squares	Surface contamination? or Surface charging?
Gradual decrease in a selected x-ray spectrum	Surface charging? or Beam damage?
Decrease in the Duane-Hunt limit	Surface charging of the whole x-ray spectrum

A similar list is shown below of image aberrations and artifacts that may be seen as a consequence of using faulty charge elimination procedure which occur at the end of specimen preparation (Table 12.2).

Most damage and artifacts are found in organic polymers and biological samples because they frequently need the use of invasive chemical stabilization prior to microscopy and analysis and are more readily damaged by the electron beam.

The recent book by Sawyer et al. (2008) contains details of damage and artifacts in polymers. Figures 12.1 and 12.2 show the types of damage and contamination in polymers.

The books by Crang and Klomparens (1988), Hayat (2000), Bozzola and Russell (2000), and Maunsbach and Afzelius (1999)

Figure 12.1. The effect of beam damage on SE images of the surface of a molded poly(methylmethacrylate) PMMA specimen. **A.** The first image taken of the specimen. **B.** A later image taken at lower magnification than image *(A)*. **C.** A higher magnification image taken *after* taking image *(B)*. The horizontal lines across images *(B)* and *(C)* are caused by the electron beam. However, these horizontal lines could be cracking of a thick metal coating layer or a genuine sample defect if only image *(C)* was examined (Picture courtesy of Sawyer et al., 2008)

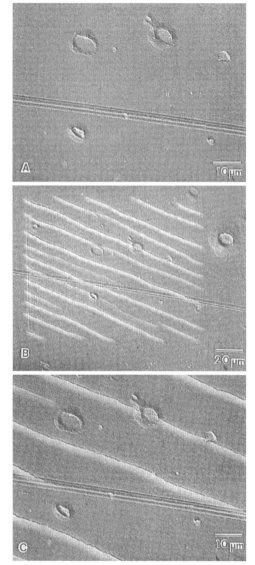

contain useful guides to recognize, identify, and understand the problems which occur in biological specimens. In addition, many published papers draw attention to damaged image and suggest ways of overcoming these problems.

Figures 12.3 and 12.4 show a number of images which the author took 40 years ago. They all show different forms of damage.

Sometimes the damage is much less obvious, for although the specimen has probably been adequately prepared, the primary electron beam has caused subtle changes to the image as shown in Figure 12.5.

An interesting series of artifacts arose during the early studies on low temperature scanning electron microscopy and analysis

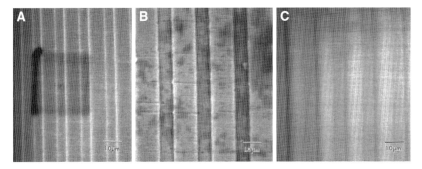

Figure 12.2. SE images of a polycarbonate grooved optical disc. **A.** Image at 1 keV shows dark contamination in the picture area previously examined at higher magnification. **B.** Image at 3 keV shows mottled surface contamination and a fine linear pattern. **C.** Image at 1 keV shows fine horizontal patterns. The horizontal and linear patterns are due to faulty setting of the field emission source (Pictures courtesy of Sawyer et al., 2008)

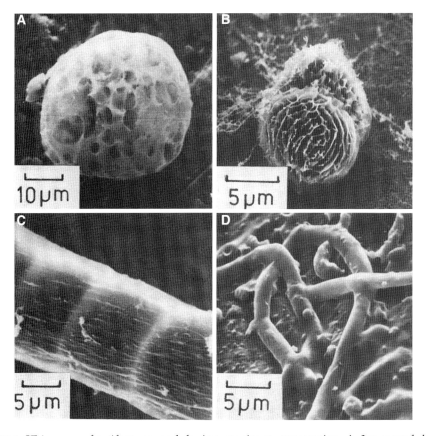

Figure 12.3. SE images of artifacts caused during specimen preparation. **A.** Ice crystal damage in a badly freeze-dried Clarkia pollen grain. **B.** Incomplete removal of surface mucous material during specimen stabilization of the single cell alga, Chroococcus. **C.** Partial collapse of a poorly critical point dried alga Lyngbia. **D.** Incomplete cleaning of the slime layer found on the surface of the bacteria *Bacillus megatherium*. All image taken at 10 keV from specimens coated with 8 nm gold

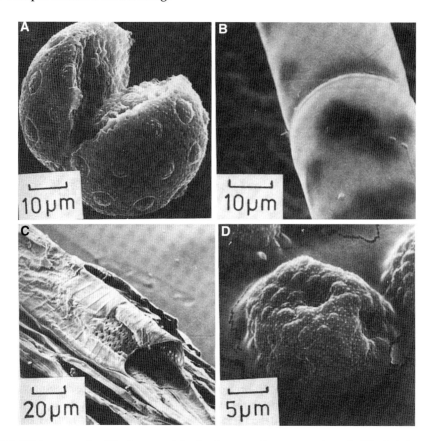

Figure 12.4. SE images of artifacts caused by faulty sample preparation. **A.** The bright area reveal incomplete metal surface coating of a freeze fractures and freeze dried *Silene* pollen grain. **B.** Dark areas of contamination on the surface of a stabilized and metal coated filamentous alga *Oedogonium*. **C.** Collapsed xylem vessels from fresh *Tulipa* stems that had been poorly dried. **D.** A partially collapsed *Plantago* pollen grain that had sunk into the polymer mounting material. Images coated with 8 nm gold and examined at 10 keV

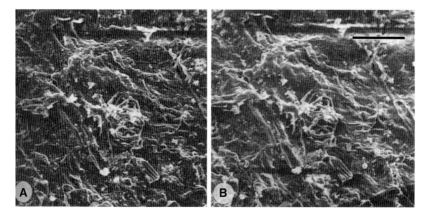

Figure 12.5. A 200-μm thick section of tobacco sheet material containing 12% water, quench cooled at 77 K, sputter coated with 6 nm of chromium and maintained at 110 K on the SEM cold stage. **A.** SEM image of the sample digitally scanned at a dose of 6.25 ke$^-$ nm^{-2} per pixel. **B.** The same region scanned at a dose of 3.13 ke$^-$ nm^{-2}. Image *(A)*, which was scanned after Image b, shows less fine detail and some beam damage

Figure 12.6. The different images of the same 500 nm cryo-section of a quench cooled 10% aqueous solution of serum albumin examined in four different ways in an SEM. **A.** SE image of frozen-hydrated material. **B.** Transmitted electron (TE) image of frozen-hydrated material. **C.** SE image of frozen-dried material. **D.** TE image of frozen-dried material. The two frozen hydrated images were first taken at 153 K and then the sample allowed to slowly freeze dry in the microscope column. The two frozen-dried images were taken at 300 K. All the photographs were taken at 5 keV and 1 nA beam current of uncoated material. Bar marker = 10 μm. Note the virtual reversal of contrast between the SE and TE images and the frozen-hydrated and freeze-dried image. The surface contaminating ice disappeared when the sample was freeze-dried

(see Echlin 1992 for details). Figure 12.6 show four different images of the same sample imaged in four different ways. Once the significance of these artifacts was understood they have provided useful diagnostic information.

The actual interpretation of the image obtained by the SEM is, paradoxically, both the easiest and the most difficult part of the whole process. It is easy, simply because the images are familiar to us as we live in a world that is illuminated from above and we are blessed with stereoscopic vision.

It is difficult, particularly when a new feature first appears as a high resolution image. An interpretation involves examining similar well characterized sample at varying magnifications.

The only advice one can offer to an individual experimental microscopist is to assemble a black list of images of damaged

Table 12.3. A suggested operational pathway to follow when trying to recognize damage to specimen images and analytical data and understand the significance of artifacts

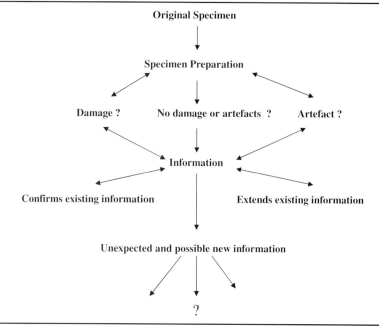

samples and of images that may have a spurious interpretation. Each image should be accompanied by the full details of how the sample was prepared (and examined!). This black book should be kept for future reference. In many cases damage and artifact can only be understood by comparing the prepared sample with an identical unprepared sample at a range of magnifications from the resolution of the human eye to an electron beam instrument.

Table 12.3 is a suggested pathway to follow when trying to untangle the causes of specimen damage and appreciate and judge the significance of artifacts. In most cases the pathways will be bi-directional, as the preparative procedures and operation of the SEM are repeated and improved. The end point of the single uni-directional pathway is the most intriguing for it may well herald some genuinely new information and it is here that serendipity can come in to play.

Serendipity has many different explanations, but in essence it is the facility to make chance discoveries accidentally that Louis Pasteur said are only favored by the prepared mind. Examples of serendipity in science are seen in the discovery of Teflon, Viagra, the smallpox vaccine, and high temperature superconductors.

13
Additional Sources of Information

13.1 Introduction

The previous 12 chapters have taken a broad overview of the principles behind the various steps that need to be taken to prepare specimens for microscopy and analysis in a scanning electron microscope. It is impossible to cover *every* aspect of *every* type of sample and this final chapter provides a guide to additional and more specific information. The first part of this chapter indicates where to find additional information, followed by a list of places where one may purchase the tools, equipment, and consumable supplies needed to carry out sample preparation.

13.2 Verbal Information

13.2.1 International Societies

The International Federation of Societies of Microscopy (IFSM) (www.ifsm.umn.edu), made up of about 40 different societies from all over the world, holds a large meeting every 4 years in different countries. IFSM sponsors regional meetings in the intervening years. These are very large meetings, which attract a large number of delegates and have a large and comprehensive commercial display of goods and equipment. Abstracts of the presented papers are published as hard copy proceedings and, recently, as a CD-ROM.

13.2.2 Scientific Societies

Most developed and some developing countries have active microscope societies that include details of both scanning electron microscopy and x-ray microanalysis. These societies have both general and specific subject meetings and publications that cover all aspects of the subject. Membership of such societies is not expensive and readers should consider applying to join their local groups that are valuable sources of information.

The author of this book is a member of the two principal English language societies, the Royal Microscopical Society (RMS) (www.rms.org.uk) and the Microscopy Society of America (MSA) (www.msa.microscopy.com).

The RMS has a biennial MicroScience meeting in London and usually has some sessions on SEM and microanalysis. Short (100-word) abstracts are published of all the presentations and selected papers are published in the *Journal of Microscopy* (see the following). A commercial exhibition displays the goods and services associated with microscopy.

The MSA sponsors the annual Microscopy & Microanalysis conference in different parts of the United States each August, usually with participation from one or more microscopy societies from North America, South America, or the Pacific Rim countries. These annual conferences typically attract nearly 3,000 participants and the 800 or so presented papers are published as two-page abstracts, either in hard copy or on a CD-ROM. This meeting is the largest annual microscopy conference in the world. As part of the meeting, an extensive commercial exhibition displays the goods and services associated with microscopy in all its forms, including specimen preparation equipment for SEM, TEM, and x-ray microanalysis.

Similar societies are found in many different countries and their information is available in their native language and, frequently, in English. In addition to the microscope societies, there are many different general and specific subject societies that include information about scanning electron microscopy and x-ray microanalysis. This information is readily available by using one of the search engines available on modern computers.

13.2.3 Courses and Workshops

A number of individuals, commercial organizations, universities, and scientific societies run courses and workshops on scanning electron microscopy and x-ray microanalysis. Probably the best-known English language course is the Lehigh Microscopy School (www.lehigh.edu/microscopy), which has been running annually in early June for nearly 40 years. The McCrone Research Institute www.mcri.org also runs courses on many aspects of microscopy. Small specialist meetings are organized by microscope societies.

13.3 Written Information

A large number of references to specimen preparation are already scattered through the chapters of this book. A few of these references are very general, but most are specific for a particular aspect of specimen preparation. The author has found three additional references very useful when considering the specimen preparation needed for imaging and analysis in the SEM.

1. The *Handbook of Chemistry and Physics* published by CRC Inc. (www.crcpress.com). This is a massive, 2,600 page book, full of accurate, reliable, and up-to-date information of the physical and chemical properties of inorganic and organic compounds and materials. The 88th edition of this book was published in 2007.

2. The *Merck Index* 14th Edition 2006 is available from scientific book shops. This is an encyclopedia of chemicals, drugs, and biological compounds. It is also available as a CD-ROM.

3. "Current Contents" is a multidisciplinary web resource that covers about 8,000 publications. It is available electronically and as a hard copy publication. www.scientific.thompsonreuters.com.

13.4 General Journals of Microscopy and Analysis

The following list is of some of the more prominent English language journals that cover general microscopy and analysis that may contain information relating to specimen preparation. A number of the journals are the official publication of different microscope societies or affiliation with microscope societies.

Acta Microscopica www.actamicroscopica.ivic.ve
Infocus www.rms.org.uk
Imaging and Microscopy www.wiley.com
Journal of Electron Microscopy www.oup.co.uk
Journal of Electron Microscopy Techniques www.wiley.com
Journal of Microscopy www.blackwellpublishing.com
Microbeam Analysis Society Proceedings www.microbeamanalysis.org/masjh/sh.php
Micron www.elsevier.com/locate/micron
Microscopy and Microanalysis www.cambridge.org
Microscopy Research and Techniques wwwinterscience.wiley.com
Microscopy Today www.microscopy-today.com
Scanning www.interscience.wiley.com
SEM Proceedings, 1968-1977 www.iitri.org 1978-www.rsc.org
Ultramicroscopy www.elsevier.com/locate/ultramicroscopy

Although it is invidious to highlight one of these very informative publications, the bi-monthly *Microscopy Today* published by the MSA contains a section called Netnotes, which is full of useful information about specimen preparation. This resource is available as a Listserver (listserver@msa.microscopy.com) to which questions may be posted. The questions may be read by members of the MSA and, if appropriate, the answers are published in the Netnotes section of *Microscopy Today*.

13.5 Specific Discipline Journals Containing Specimen Preparation Methods

After consulting colleagues from all over the world and spending time in the many different libraries in the University of Cambridge, a list of publications emerges that usually contains one or more papers about methods of specimen preparation for each of the six groups that have been considered in this book.

13.5.1 Metals, Alloys, and Metallic Materials

Acta Materials www.elsevier.com/locate/actamat
Corrosion Science www.elsevier.com/locate/corsci
Intermetallics www.elsevier.com/locate/intermet
Journal of Electronic Materials www.springer.com
Journal of Materials Research www.mrs.com
Journal of Materials Science www.springer.com
Materials Science and Engineering www.elsevier.com/locate/msea
Metallurgy and Materials www.asm.com
Metallurgical Transactions www.tms.com
Powder Metallurgy www.maney.co.uk

13.5.2 Hard, Dry Inorganic Materials

Advances in Applied Ceramics www.maney.co.uk
American Mineralogist www.minsocam.org
American Association of Petrolium Geologists Bulletin www.aapg.org
Earth and Planetery Science Letters www.elsevier.com
European Journal of Mineralogy www.schweizerbart.de
Geochimica et Cosmochica Acta www.elsevier.com/locate/gca
Geological Magazine www.cambridge.org
Geological Society of America Bulletin www.geosociety.org
Journal of the Clay Minerals Society www.clays.org
Journal of Soil Science www.elsevier.com
Journal of the Geological Society www.geolsoc.org.uk
Meteoritics and Planetary Science www.metbase.de
Minerology and Petrology www.springer.at.minpet
Palaeontology www.palass.org

13.5.3 Hard or Firm, Dry Natural Organic Materials

Journal of Wood Chemistry and Technology www.tandf.co.uk
Journal of Pulp and Paper Science www.tappi.org

13.5.4 Synthetic Organic Polymer Material

European Polymer Journal www.elsevier.com/locate/europoly
Journal of Applied Polymer Science www.elsevier.com
Polymer www.elsevier.com/locate/polymer

13.5.5 Biological Organisms and Materials

Acta Biomaterials www.elsevier.com/locate/actabiomat
American Journal of Botany www.amjbot.org
Annals of Botany www.aob.oxfordjournals.org
Cell Biology www.elsevier.com/locate/cellbio
Journal of Biotechnology www.elsevier.com/locate/jbiotech
Journal of Food Engineering www.elsevier.com/locate/foodeng
Journal of Food Science www.blackwellpublishing.com
Journal of Cell Biology www.biologists.org
Microbiology www.sgm.ac.uk
Planta www.springer.com
Plant Science www.elsevier.com/locate/plantsci
The Plant Cell www.plantcell.org

13.5.6 Wet and Liquid Samples

European Journal of Pharmaceutics and Technology www.elsevier.com
Journal of Colloid and Interface Science www.elsevier.com
Journal of Industrial Microbiology and Biotechnology www.springer.com

13.6 Publication Listing Consumable Supplies, Equipment, and Tools Needed for Sample Preparation

Although much of this type of information is located in different chapters of this book, it is convenient to attempt to consolidate this material. Large pieces of equipment such as scanning microscopes, x-ray analytical spectrometers, and detectors, vacuum equipment, critical point dryers, evaporative and sputter coaters, and thin film coating devices will not be included. Additional and continuing information has been culled from the annual *Expo Issue of Microscopy and Microanalysis* (www.journals.cambridge.org), the annual *Supplement to Microscopy and Analysis* (www.microscopy-analysis.com) and from trade publications. The latest issue of the *UK-European Microscopy Accessories Guide* was published in June 2008. Similar issues for the Americas and for Asia are published annually.

13.6.1 Specific Sources of Consumable Supplies, Equipment and Tools Needed for Sample Preparation

13.6.1.1 Accessories (Miscellaneous)

ETS-Lindgren www.ets-lindgren.com
Halcyonics Inc. www.halcyonics.com
Quantomix www.quantomix.com
Small World www.small-worrld.com

13.6.1.2 Chemicals

AgarScientific www.agarscientific.com
Buehler www.buehler.com

Cargille Laabd www.cargille.com
Diatome U.S. www.emsdiasum.com
Electron Microscopy Sciences www.emsdiasum.com
Ernest F Fullam Inc. www.fullam.com
Ladd Research www.laddresearch.com
Polysciences www.polysciences.com
SPI Supplies www.2spi.com
Taab www.aab.co.uk

13.6.1.3 Consumable Supplies (General)

AgarScientific www.agarscientific.com
Allied High Tech products www.alliedhightech.com
Ascen Instruments www.ascendinstruments.com
Duniway Stockroom, Corp. www.duniway.com
Earnest F Fuller Inc. www.fullam.com
E.A.Fischione Instruments Inc. www.fischione.com
Electron Microscopy Sciences www.emsdiasum.com
Emitech Ltd. www.emitech.co.uk
Ladd Research www.laddresearch.com
Mager Scientific Inc. www.magersci.com
McCrone Microscope & Accessories www.mccrone.com
Pelco International www.pelcoint.com
Refining System Inc. www.refyningsystems.com
Technotrade International Inc. www.technotradeinc.com
Triangle Biomedical Services Inc. www.trianglebiomedical.com

13.6.1.4 Cryo-equipment

Attocube Systems AG www.attocube.com
BalTec/RMC www.baltec-RMC.com
Boeckeler Instruments Inc. www.boeckeler.com
Diatome U.S. www.emsdiasum.com
Energy Beam Sciences Inc. www.ebsciences.com
Leica Microsystems Inc. www.leica-microsystems.com
Micro Star Technologies www.microstartech.com

13.6.1.5 Cutting, Grinding, Polishing, Sawing, and Thinning

AgarScientific www.agarscientific.com
Allied High Tech Products www.alliedhightech.com
Atomic Force F&F www.atomicforce.de
Bueher Ltd. www.buehler.com
E.A.Fischione Instruments Inc. www.fischione.com
Electron Microscopy Sciences www.emsdiasum.com
Gatan www.gatan.com
South Bay Technology Inc www.southbaytech.com
Struers Ltd. www.struers.com
Techoorgg Linda www.alicona.com
Ted Pella Inc. www.tedpella.com
Ultra Tec www.ultratecusa.com

13.6.1.6 Data Storage

Kodak Molecular Imaging System www.kodak.com
Lumenae Corporation www.lumenare.com
Miero Inc. www.miero.com
Olympus Soft Imaging Solutions www.soft-imaging.net
Optomics www.optomics.com

13.6.1.7 Diamond Knives

Delaware Diamond Knives Inc. www.ddk.com
Diatome U.S. www.emsdiasum.com
Ladd Research www.laddresearch.com
Micro Star Technologies www.microstartech.com

13.6.1.8 Histochemical Equipment and Supplies

Diatome U.S. www.emsdiasum.com
Electron Microscopy Supplies, www.emsdiasum.com
Micros Produktions www.meijitecho.com
Molecular Probes www.probes.com
SPI Supplies www.2spi.com
Ted Pella Inc. www.tedpella.com

13.6.1.9 Knives

AgarScientific www.agarscientific.com
Delaware Diamond Knives Inc. www.ddk.com
Diatome U.S. www.emsdiasum.com
ISI Group www.iss-group.co.uk
Taab www.taab.co.uk

13.6.1.10 Knife Resharpening

Delaware Diamond Knives Inc. www.ddk.com
Diatome U.S. www.emsdiasum.com
Dorn & Hart Microedge www dornandhart.com
Micro Star Technologies www.microstartech.com

13.6.1.11 Metallographic Equipment

Alicona Imaging GmbH www.alicona.com
Allied High Tech Products www.alliedhightech.com
Buehler www.buehler.com
Ladx www.ladx.com
Princeton Gamma-Tech Instruments Inc. www.pgt.com
Qualitest Inc. www.worldoftest.com
South Bay Technology Inc. www.southbaytech.com

13.6.1.12 Micromanipulators and Precision Tools

Kleindiek Nanotechnik GmbH www.nanotecnik.com
Micro Optics of Florida www.microopticsfl.com
Minotool Inc. www.minitoolinc.com

13.6.1.13 Microtomes and Ultramicrotomes

BAL-TEC/RNC www.baltec-RMC.com
Boeckeler Instruments Inc. www.boeckeler.com
Leica Microsystems Inc. www.leica-microsystems.com
Mager Scientific Inc. www.magersci.com
Micro Star Technologies www.microstartech.com
Triangle Biomedical Services www.trianglebiomedical.com

13.6.1.14 Microwave Tissue Processing

Energy Beam Science Inc. www.ebsciencs.com
Ted Pella Inc. www.tedpella.com

13.6.1.15 Photographic and Imaging Supplies

Earnest F Fuller Inc. www.fullam.com
Fuji Photo Films USA www.fugifilm.com
Ladd Research www.laddresearch.com
Photron USA www.photron.com

13.6.1.16 Reference Materials for Analysis and Calibration

Energy Beam Science Inc. www.ebsciencs.com
Ted Pella Inc. www.tedpella.com
Geller MicroAnalytical Laboratory www.gellermicro.com

13.6.1.17 Small Tools and Blades

Diatome U.S. www.emsdiasum.com
Geller MicroAnalytical Laboratory www.gellermicro.com
Minitool www.minitoolsinc.com
Ted Pella Inc. www.tedpella.com

13.6.1.18 Thin Foils, Films, Grids, Metal, Materials, Supports, and Meshes

Agar Scientific www.agarscientific.com
Energy Beam Sciences Inc. www.ebsciences.com
Goodfellow Metals www.goodfellow-metals.co.uk
Physical Electronic www.phi.com
Polysciences www.polysciences.com
Protochips Inc. www.protochips.com
Refining Systems Inc. www.refiningsystems.com
SPI Supplies www.2spi.com
Ted Pella Inc. www.tedpella.com

13.6.1.19 X-ray Analysis Supplies

Geller MicroAnalytical Laboratory www.gellermicro.com
Norrox Scientific Ltd. www.mag-i-cal.ca
Refining Systems Inc. www.refiningsystems.com
Richard Technologies www.richardson-tech.com
Ted Pella Inc. www.tedpella.com

References

Aceterin, J.-D., Carlemalm, E., and Villiger, W. (1986) J. Microsc. 143, 81.
Aceterin, J.D., Carlemalm, E., Kellenberger, E., and Villiger, W. (1987) J. Electron. Microsc. Technol. 6, 63.
Akahori, H., Handa, M., Yoshida, H., and Kozuka, Y. (2000) J. Electron. Microsc. 49, 735.
Albright, R.M. and Lewis, R. (1965) J. Appl. Phys. 36, 2615.
American Society of Metals, Metals Park, Ohio 44073.
Anderson, T.F. (1951) Trans. N Y Acad. Sci. 13, 13.
Armbruster, B.L., Carlemalm, E., Chivetti, R., Caravito, R.M., Hoboy, R.H., Kellenberger, E., and Villiger, W. (1982) J. Microsc. 126, 77.
Bancroft, J.D. and Cook, H.C. (1984) Manual of histochemical techniques. Churchill Livingstone, Oxford.
Bancroft, J.D. and Stevens, A. (1977) Theory and practice of histochemical techniques. Churchill Livingstone, Oxford.
Barnard, T. and Thomas, R.S. (1978) J. Microsc. 112, 281.
Bencosme, S.A. and Tsutsumi, V. (1970) Lab. Invest. 23, 447.
Boggs, S. and Krinsley, D. (2006) Application of cathodoluminescence imaging to the study of sedimentary rocks. Cambridge University Press, New York.
Bousfield, B. (1992) Surface preparation and microscopy of materials. Wiley, New York.
Bowling, A.Y. and Vaughn, K.C. (2008) J. Microsc. 231, 186.
Boyde, A., Bailey, E., Jones, S.J., and Tamarin, A. (1977) SEM 1, 507.
Boyde, A. and Maconnachie, E. (1979) Scanning 2, 149.
Boyde, A. (2004) Adv. Imag. Electron Phys. 133, 165.
Bozzola, J.J. and Russell, L.D. (1999) Electron microscopy. Principles and techniques for biologists. Jones & Bartlett, Sudbury, MA.
Breton, B.C., McMullan, D., and Smith, K.C.A. (eds.) (2004) Sir Charles Oatley and scanning electron microscopy. Vol. 133. Advances in imaging and electron physics. Elsevier, Amsterdam.
Bridgeman, P.C. and Reese, T.S. (1984) J. Cell Biol. 99, 1655.
Buehler Ltd. (1973) Petrographic sample preparation. Lake Bluff, IL.
Bullock, G.R. (1984) J. Microsc. 133, 1.
Carlemalm, E., Garavito, R.M., and Villiger, W. (1982) J. Microsc. 126, 123.
Carlemalm, E. and Villiger, W. (1989) Tech. Immunocytochem. 4, 29.

Carlemalm, E., Villiger, W., Hobot, J.A., Aceteran, J.D., and Kellenberger, E. (1985) J. Microsc. 140, 55.
Chang, C.C.Y. and Alexander, J.W. (1981) Biol. Cell 40, 99.
Chen, Y., Centronze, V.E., Verkhovsky, A., and Borisy, G.G. (1995) J. Microsc. 179, 67.
Cizek, P., Wynne, B.P., and Rainforth, W.M. (2006) J. Microsc. 222, 85.
Clay, C.S. and Peace, G.W. (1981) J. Microsc. 123, 25.
Crang, R.F.E. and Klomparens, K.L. (1988) Artefacts in biological microscopy. Plenum, New York.
Crissman, R.S. and McCann, P. (1979) Micron 10, 37.
Dabbs, D. (2006) Diagnostic immunocytochemistry. Churchill Livingstone, Oxford.
Dashek, W.V. (ed.) (2000) Plant electron microscopy and cytochemistry. Humana, Totowa, NJ.
Davis, T.W. and Morgan, A.J. (1976) J. Microsc. 107, 47.
Douchet, R.G. and Bradley, S.A. (1989) J. Electron. Microsc. 12, 58.
Douzou, P. (1997) Cryochemistry. Academic, New York.
Echlin, P. (1992) Low temperature microscopy and analysis. Plenum, New York.
Echlin, P. (1996) Scanning 18, 197.
Echlin, P. (2004) Adv. Imag. Electron. Phys. 133, 469.
Echlin, P. and Taylor, S.E. (1986) J. Microsc. 141, 329.
Edwards, H.H., Yeh, Y.'Y., Tarmow, B.I., and Schonbaum, G.B. (1992) Microsc. Res. Technol. 21, 29.
Ellis, E.A. (2006) Microsc. Today 14, 50.
Ellis, E.A. and Pendleton, M.W. (2007) Microsc. Today 13, 44.
Ellisman, M.H.P., Friedmann, L., and Hamillton, W.J. (1980). J. Neurosci. 9, 185.
Ensikat, H.J. and Barthlott, W. (1993) J. Microsc. 172, 195.
Erlandsen, S., Chen, Y., Frethem, C., Detry, J., and Wells, C. (2003) J. Microsc. 211, 212.
Fernandez-Moran, H. (1959) Adv. Electron. Phys. Suppl. 16, 167.
Fisher, K.A. (1982) Methods Enzymol. 88, 230.
Flood, P.R. (1980) SEM/1980 1, 183.
Franks, F. (1985) Biophysics and biochemistry at low temperatures. Cambridge University Press, Cambridge.
Franks, F. (2007) Freeze-drying of pharmaceuticals and biopharmaceuticals—theory and practice. Royal Society of Chemistry, Cambridge.
Friel, J.J. and Lyman, C.E. (2006) Microsc. Microanal. 12, 2.
Gamliel, H. (1985a) SEM/1985 2, 929.
Gamliel, H. (1985b) SEM/1985 4, 1649.
Ge, H., Suszynski, W.J., Davis, H.T., and Scriven, L.E. (2008) J. Microsc. 229, 115.
Gerrits, P.O., Eppinger, B., Van Goor, M., and Horobin, R.W. (1991) Cell Mater. 1, 189.
Gerrits, P.O. and Horobin, R.W. (1996) J. Histotechnol. 19, 297–311.
Geritts, P.O., Horobin, R.W., and Hardonk, M.J. (1990) Histochem. J. 22, 439.
Gerrits, P.O. and van Leeuwen, M.B.M. (1985) J. Microsc. 139, 303.
Gerrits, P.O., van Leeuwen, M.B.M., and Boon, M.E. (1987) J. Microsc. 145, 107.
Giannuzzi, L.A. and Stevie, F.A. (2005) Introduction to focused ion beams. Springer, New York.
Giberson, R.T. and Demaree, R.S. (2001) Microwave techniques and protocols. Humana, Totowa, NJ.

Glauert, A.M. (1991) Microsc. Anal., September 1991, 15–20.
Glauert, A.M. and Lewis, P.R. (1998) Biological specimen preparation for transmissiom electron microscopy. In Practical methods in electron microscopy (A.M. Glauert, ed.), Vol. 17. Portland, London.
Glauert, A.M., Rogers, G.E., and Glauert, R.H. (1956) Nature 178, 803.
Goldstein, J., Newbury, D., Echlin, P., Joy, D., Romig, A.D., Lyman, C., Fiori, C., and Lifshin, E. (1992) Scanning electron microscopy and x-ray microanalysis. Plenum Press, New York.
Goldstein, J., Newbury, D., Joy, D., Lyman, C., Echlin, P., Lifshin, E., Sawyer, L., and Michael, J. (2004) Scanning electron microscopy and x-ray microanalysis. Springer, New York.
Goldstein, J.I., Jones, R.H., Kotula, P.G., and Michael, J.R. (2007) Meteorit. Planet. Sci. 42, 1.
Goodhew, P.J. (1973) Specimen preparation in material science. In Practical methods in electron microscopy (A.M. Glauert, ed.), Vol. 1. North-Holland/American Elsevier.
Goodhew, P.J. (1985) Thin foil preparation for electron microscopy. In Practical methods in electron microscopy (A.M. Glauert, ed.), Vol. 11. Elsevier, Amsterdam.
Griffiths, G. (1993) Fine structure immunocytochemistry. Springer, Berlin.
Gupta, P.D. (ed.) (2000) Electron microscopy in biology and medicine. Whitaker, House Inc, New Kensington PA.
Hariharan, H., Koschan, A., Bidi, B., Page, D., Abidi, M., Frafjord, J., and Dekanich, S. (2008) Microsc. Today March 2008, 18.
Hawes, P., Netherton, C.L., Mueller, M., Wileman, T., and Monaghan, P. (2007) J. Microsc. 226, 182.
Hayat, M.A. (1975) Positive staining for electron microscopy. Van Nostrand Reinhold, New York.
Hayat, M.A. (2000) Principles and techniques of electron microscopy. Biological applications, 4th edition. Cambridge University Press, Cambridge.
Hayat, M.A. (2002) Microscopy, immunocytochemistry, and antigen retrieval methods: For light and electron microscopy. Springer, New York.
Hayles, M.F., Stokes, D.J., Phifer, D., and Findlay, K.C. (2007) J. Microsc. 226, 263.
Hermann, R.J., Pawley, J., Nagatani, T., and Muller, M. (1988) Scan. Microsc. 2, 1215.
Heuser, J.E. (1989) J. Electron. Microsc. Technol. 13, 224.
Hirschberg, R.M., Mulling, C.K.V., and Bragulla, H. (1999) Microsc. Res. Tech. 45, 184.
Hohenberg, H., Tobler, M., and Muller, M. (1996) J. Microsc. 183, 133.
Hohenberg, H., Muller-Reichert, T., Schwarz, H., and Zierold, K. (2003) J. Microsc. 212, 3.
Holman, W.R. (1974) In Principles and techniques for electron microscopy (M.A. Hayat, ed.), Chapter 7 in Vol. 4. Van Nostrand Reinold, New York.
Holt, D.H. and Joy, D.C. (eds.) (1989) SEM microcharacterization of semiconductors. Academic, New York.
Hunt, C.J. (1984) Isotherrma; freeze fixation. In Science of biological sample preparation (J.P. Revel, T. Barnard, and G. Haggis, eds.), p. 123. SEM, Chicago, IL.
Hunziker, E.B. and Schenk, R.K. (1984) Cryo-methods for transmission electron microscopy of calcified cartilage. In Methods of calcified tissue preparation (G.R. Dickson, ed.), p. 199. Elsevier, Amsterdam.
Ingram, M.J. and Hogben, C.A. (1967) Anal. Biochem. 18, 45.
Isabell, T.C., Fischione, P.E., O'Keefe, C., Guruz, M.U., and Dravid, V.P. (1999) Microsc. Microanal. 5, 126.

Johansen, B.V. and Namork, E. (1984) J. Microsc. 133, 8.
Joy, D.A. and Joy, C.S. (1996) Micron 27, 247.
Kellenberger, E., Carlemalm, E., and Villiger, W. (1986) In Science of biological specimen preparation (M. Muller, R.P. Becker, A. Boyde, and J.L. Wolosewick, eds.), p. 147. SEM, Chicago, IL.
Kellenberger, E., Carlemalm, E., Villiger, W., Roth, J., and Caravito, R. (1980) Low denaturation embedding for electron microscopy of thin sections. Chemisch Werke Lowi GmbH. Postfach.1660, D-8266 Waldkaiburg, Germany.
Kim, Y.-N., Kang, J.-S., Kim, J.-S., Jeung, J.-M., Lee, J.-Y., and Kim, Y.-J. (2007) Microsc. Microanal. 13, 285.
Knoll, M. (1941) Phys. Z. 42, 120.
Krishna, S. (ed.) (2001) Handbook of thin film deposition, processing and techniques, 2nd edition. William Andrews, Norwich, NY.
Lamb, J.C. and Ingram, P. (1979) SEM/1979 2, 459.
Lane, W.C. (1970) Proceedings SEM Symposium (O. Johari, ed.), p. 43. IITRI, Chicago, IL.
Läuchli, A., Stelzer, R., Guggenheim, R., and Henning, L. (1978) Preparation techniques as a means of intracellular ion localization by use of electron probe analysis. In Microprobe analysis as applied to cells and tissues (T.A. Hall, P. Echlin, and R. Kaufmann, eds.), p. 109. Academic, New York.
Lee, W.E. and Rainforth, W.M. (1994) Ceramic microstructures: Property control by processing. Chapman & Hall, New York.
Leipins, A. and de Harven, E. (1978) SEM/1978 2, 37.
Lewis, P.R. and Knight, D.P. (1992) Staining methods for sectioned materials. In Practical methods in electron microscopy (A.M. Glauert, ed.), Vol. 5. North Holland, Amsterdam.
Lieberman, M.A. and Lichlerberg, A.J. (2005) Principles of plasma discharge and materials processing. Wiley, New York.
Locquin, M.W. and Langeron, M. (1978) Handbook of microscopy. Butterworths, London.
Login, G.R. and Dvorak, A.M. (1994) The microwave toolbook. Beth Israel Corporation, Boston, MA.
Luft, J.H. (1961) J. Biophys. Biochem. Cytol. 9, 409.
MacRae, C.H. and Wilson, N.C. (2008) Microsc. Anal. 14, 184.
Maisall, L.I. and Glang, R. (eds.) (1970) Handbook of thin film technology. McGraw Hill, New York.
Marshall, A.T. and Carde, D. (1984) J. Microsc. 134, 113.
Marshall, A.T., Carde, D., and Kent, M. (1985) J. Microsc. 139, 335.
Massover, W.H. (2008) Microsc. Microanal. 14, 126.
Maunsbach, A.B. and Afzelius, B.A. (1999) Biolomedical electron microscopy. Academic, New York.
Mazia, D., Sale, W.S., and Schatten, G. (1974) J. Cell Biol. 63, 212.
McDonald, K. and Muller-Reichert, T. (2008) J. Microsc. 230, 230.
McDonald, K.L., Morphew, M., Verkade, P., and Muller-Reichert, T. (2007) Methods Mol. Biol. 369, 143.
McGeoch, J.E.M. (2007) J. Microsc. 227, 172.
Michael, J.R. (2000) In Electron backscatter Diffraction in Materials Science (A.J. Schwartz, M. Kumar, and B.L. Adams, eds.). Kluwer/Academic, New York.
Morgan, A.J. (1980) Preparation of specimens: Changes in chemical integrity. Chapter 2. In X-ray microanalysis in biology (M.A. Hayat, ed.). University Park Press, Baltimore, MD.
Morgan, A.J. and Davies, T.W. (1982) J. Microsc. 125, 103.

Morgan, A.J., Davies, W., and Erasmus, D.A. (1978) Specimen preparation. In Electron probe microanalysis in biology (D.E. Erasmus, ed.). Chapman & Hall, London.
Muller, L.L. and Jacks, T.J. (1975) J. Histochem. Cytochem. 23, 107.
Murakami, T. (1978) Scanning 1, 127.
Murakami, T., Yamamoto, K., and Itoshima, T. (1977) Arch. Histol. Jpn. 40, 35.
Murakami, T.H., Iia, O., Tagechi, A., Ohtami, A., Kikuta, A., Ohtsuka, A., and Itoshima, T. (1983) Scan. Electron Microsc. 1, 235.
Murphy, J.A. (1978) Scan. Electron Microsc. 2, 175.
Nation, J.L. (1983) Stain Technol. 58, 347.
Newman, G.R. and Hobot, J.A. (1993) Resin microscopy and on-section immunocytochemistry. Springer, Berlin.
Oatley, C.W. (1972) The scanning electron microscope part 1. The instrument. Cambridge University Press, Cambridge.
Osawa, T., Yoshida, F., Tsuzuku, T., Nozaku, M., Takashio, M., and Nozaku, Y. (1999) J. Electron. Microsc. 48, 665.
Paulson, G.G. and Pierce, R.W. (1972) Proceedings of 30th Annual EMSA. Claiters, Baton Rouge, LA, p. 406.
Pearse, A.G.E. (1985) Histochemistry. Theoretical and applied. Vol. 2 Analytical techniques. Churchill Livingstone, Oxford.
Peters, K.-R. (1980) SEM/1980 1, 143.
Peters, K.-R. (1985) SEM/1985 4, 1519.
Peters, K.-R. (1986) J. Microsc. 142, 25.
Petzow, G. (1978) Metallographic etching. American Society of Materials, Metal Park, Ohio 4407.
Polak, J.M. and Van Noorden, S. (2003) Introduction to immunocytochemistry. Garlend Science, London.
Postek, M.T., Howard, K., Johnson, A.H., and McMichael, K.L. (1980) Scanning electron microscopy. A students handbook. Ladd Research Industries Inc., Williston VT 05495, USA.
Prewett, P.D. (1992) Vacuum 44, 345.
Quamme, G.A. (1988) Scan. Microsc. 2, 2195.
Randle, L. and Engler (2000) Introduction to texture analysis: Macrotexture, microtexture and orientation mapping. Gordan & Beach, New York.
Renshaw, A. (ed.) (2007) Immunohistochemistry: Methods express series. Scion, Bloxham, Oxfordshire.
Richter, T., Biel, S.S., Sattler, M., Wenck, H., Wittern, K.-P., Wiesendanger, R., and Wepf, R. (2007) J. Microsc. 225, 201.
Robard, A.W. and Sleytr, U.B. (1985) Low temperature methods in biological electron microscopy. In Practical methods in electron microscopy (A.M. Glauert, ed.), Vol. 10. Elsevier, Amsterdam.
Robin, M., Combes, R., and Rosenberg, E. (1999) CryosSEM and ESEM: New techniques to investigate phase interactions with reserve rocks. Society of Petroleum Publication 568829, Society of Petroleum Engineers, Richmond, TX.
Rosenberg, M., Bartl, P., and Lesko, J. (1960) J. Ultrastruc. Res. 4, 298.
Ryazantsev, S.N. (2000) Microsc. Anal. September 2000, 19.
Sata, S., Abachi, A., Sasaki, Y., and Chazizadeh, M. (2008) J. Microsc. 229, 17.
Sawyer, L.C., Grubb, D.T., and Meyers, G.F. (2008) Polymer microscopy, 3rd edition. Springer, New York.
Scott, R.D. (1959). ASTM Spec. Tech. Publication No. 237-121.
Sela, J. and Boyde, A. (1977) J. Microsc. 111, 229.

Shimoni, E. and Muller, M. (1998) J. Microsc. 192, 236.
Sims, P.A. and Albrecht, R.N. (1999) Microsc. Microanal. 5, 99.
Skepper, J. (2000) J. Microsc. 199, 1.
Smet, P.F., Van Haegke, J.E., and Poelmam, D. (2008) J. Miscros. 231, 1.
Smith, D.L. (1995) Thin-film deposition: Principles and practice. McGraw Hill, New York.
Smith, K.C.A. (1956) Ph.D. dissertation. University of Cambridge, Cambridge.
Spurr, S. (1969) J. Ultrastruc. Res. 26, 31.
Stenberg, M., Stemme, G., and Nygren, H. (1987) Stain Technol. 62, 231.
Stewart, A.D.G. (1962) International Conference of Electron Microscopy. Philadelphia, PA, Paper D12.
Stokroos, L., Kalicharam, D., Van der Want, J.J.L., and Jomgbloed, W.L. (1998) J. Microsc. 189, 79.
Stöttinger, B., Klein, M., Minnchich, B., and Lametschwandtner, A. (2006) Microsc. Microanal. 12, 376.
Sylvester-Bradley, C.C. (1969) Micropaleontology 15, 366.
Tanaka, K. (1989) Biol. Cell. 65, 89.
Taylor, A.P., Webb, R.I., Barry, J.C., Hosmer, H., Gould, R.J., and Wood, B.J. (2000) J. Microsc. 199, 56.
Thornley, R.F.M. (1960) Ph.D. thesis. University of Cambridge, Cambridge.
Tiedemann, J., Holenberg, H., and Kollman, R. (1998) J.Microsc. 189, 163.
Tzaphilodou, M. and Mattoupoulos, D.P. (1988) Micron. Micros. Acta. 19, 137.
Umrath, W. (1983) Mikroskopia 40, 9.
Van der Voort, G.F. (1984) Metallography: Principals and practice. McGraw Hill, New York.
Van Noorden, C.J.F. and Fredriks, W.M. (1992) Enzyme histochemistry. Oxford University Press, Oxford.
Van Steveninck, R.F.M. and Van Steveninck, M.E. (1991) Microanalysis. In Microanalysis in electron microscopy of plant cells (J.L. Hall and C. Hawes, eds.), p. 415. Academic, New York.
Villiger, W. (1991) Lowicryl resins. In Colloidal gold: Principle, methods, and applications (M.A. Hayat, ed.), p. 59. Academic, San Diego, CA.
Walter, P. (2003) Microsc. Microanal. 9, 279.
Walter, P. and Muller, M. (1999) J. Microsc. 196, 279.
Wandrol, P. (2007) J. Microsc. 227, 24.
Warley, A. (1997) X-ray microanalysis for biologists. In Practical methods in electron microscopy (A.M. Glauert, ed.), Vol. 16. Portland, London.
Wepf, R., Amrein, M., Burkli, U., and Gross, H. (1991) J. Microsc. 163, 51.
Woodward, J.T. and Zasadinski, J.A. (1996) J. Microsc. 184, 157.
Woolweber, L., Strake, R., and Gothe, U. (1981) J. Microsc. 121, 185.
Xia, J. (2001) Ph.D. thesis. University of Cambridge, Cambridge.
Yarom, R., Maunder, C., Scripps, M., Hall, T.A., and Dubowitz, V. (1975) Histochemistry 45, 59.
Zhou, W. and Wang, Z.L. (2006) Scanning microscopy for nanotechnology. Springer, New York.

Index

A

Aceterin, J. D., 72
Acrylic resins, 51, 56–61
Aerosols, 39
Afzelius, B. A., 301
Albrecht, R. N., 82
Albright, R. M., 233
Ambient-temperature wet chemical methods, 212–213
Analytical signal
   accelerating voltage, 187
   beam current, 187
Antibody–antigen interaction, 219
Araldite, 52
Auto carbon coater, for thin films production, 269

B

Backscattered electrons, 1, 137, 185, 186, 208
   annular detectors, 190
   detectors, 14
   imaging procedure, 190
   secondary electron, 209
   signals, 185, 188
Bakelite, 49–50
Barnard, T., 233
Beam damage, image artifacts, 301
Beam energy, 248
Bell jar
   clean, diagram of, 271, 272
   flushing, 275
Biological organisms and materials, 5–6
Biological samples, 11
Blades, suitable for preparing samples for SEM, 71
Bousfield, B., 48
Bowling, A. Y., 58
Boyde, A., 87, 294
Bozzola, J. J., 38, 301
Bradley, S. A., 102
Breton, B. C., 85
BSE. *See* Backscattered electrons
Bulk conductive stained rat mitochondria, 256
Bulk conductive staining techniques
   advantages/disadvantages, 257
   SEM, 258
   thermal conductivity, 255
Bulk conductivity preparative techniques, high-resolution images, 256
Burley tobacco leaves, elemental concentrations of, 229

C

Carbon, contaminating layer of, 241
Carbon film thickness, estimation of, 292
Carde, D., 284
Carlemalm, E., 132
Cathodoluminescence, 186, 194–196
   beam scatters, 194
   emission spectra, 195
   molecules/macromolecules, 194–196
   semiconductors, 195
Cell coated, 219
Charge-coupled device (CCD) camera, 14
Charging samples, contamination, 247
Chemical dehydration, 106
Chemical stabilization
   general features of, 220
   immunocytochemical localization, 220
   low temperature methods, 221
Chen, Y., 285
Chlorosulfonic acid, 141
Chromium plasma sputter, 274
CL. *See* Cathodoluminescence
Clay, C. S., 290
Clean plant specimens, 245
Coating material, removing, 293
Coating thickness
   film thickness
      calculation of, 288
      evaporative coating, 286
      quartz thin film monitor, 289
      SE images, 287
      sputter coating, 288
   SEM, 286
   x-ray microanalysis, 286
Colloidal gold probes, 220
Conducting materials, 33

Conducting tapes, 43
Conductive material, thin layers of, 258
Continuous thin film, progressive steps for formation of, 262
Copper-based microelectronic material, high resolution SEM images, 242
Corrosion
  casts, 82
  surface layers of, 204
CPD. *See* Critical point drying
Crang, R. F. E., 301
Crissman, R. S., 294
Critical point drying, 106–112
  disadvantages
    chemical integrity, 112–113
    low temperature drying, 113–114
    structural integrity, 112
CryoSEM system, 222
Cytoplasmic streaming, 211

D

Davies, W., 232
Davis, T. W., 233
De Harven, E., 102
Diallyl phthalates, 50
Diaminobenzidine (DAB), 199
2,2-Dimethoxypropane (DMP), 106
Douchet, R. G., 102
Dry inorganic samples, 204
Dry natural organic materials, 4–5
  SEM analysis, principal approaches for, 205
Dry nitrogen gas lines, vacuum pump lines, 267
Dry seeds, carbohydrates, 206
Dual beam microscopes, 298
Dual beam microscopy, 90–92
Dvorak, A. M., 156

E

Ebonite, 141–142
EBSD. *See* Electron backscattered diffraction
Echlin, P., 121, 164, 194, 228, 285, 290
EDS. *See* Energy dispersive spectrometer
EDS spectrum
  detecting charging, 251
  detecting charging on, 251
  silver specimen, 238
Electrolytic grinding, 75
Electron backscattered diffraction, 186
  BCC iron, 191
  crystallographic, 190–191
  phase information, 190–191
  SEM, 190
Electron beam, 301
    instruments
      bulk samples, 211
      isolated cells, 212
      thick sections, 212
Electro polishing, 78
Ellis, E. A., 56
Embedding, for biological samples, 48
Embedding material, criteria governing suitability of, 49
Energy dispersive analysis, frozen leaves, 228
Energy dispersive spectrometer, 192
  lower bean current, 193
Energy dispersive x-ray detectors, 14
Enzyme ATPase, plant/animal material, 215
Enzyme histochemistry, 200
Epon, 55
Epoxy resins, 50, 51
  standard infiltration schedule for, 52
  types of, 52
Erlandsen, S., 188, 219
Evaporative coating
  contamination, 295
  film adhesion, 295
  problems, 268
  thermal radiation, 294

F

Face centered cubic (FCC) metal, 68
Faulty secondary electron images, of polymers, 300
Fischione automatic sample preparation system, 241
Fizeau method, 291
Flame photometry, 214
Flood, P. R., 290
Focused ion beam (FIB) instrument, 87–88
Focused ion beam (FIB) microscope, 84
Freeze-drying, 231
  consequences of, 114–115
  damage and artifacts associated with
    analytical artifacts, 123
    molecular artifacts, 122
    structural artifacts, 122–123
  equipment for, 118–120
  guidelines for practical, 117
  liquid nitrogen cooled, 120–121
  from non-aqueous solvents, 121–122
  protocols for, 117–118
  SEM images, 222
  theoretical basis of, 115–117
Freeze-fractured root cells, SEM images of, 226
Freeze substitution, 223, 231
  advantages and disadvantages of, 131–132
  general outline of procedures used for
    specimen cooling, 126
    specimen destination, 127
    specimen stabilization, 125–126
    specimen substitution, 126
  general principles of, 124–125

low temperature refrigerators for, 127–129
practical procedures for, 129–130
  acetone, 130
  diethyl ether, 130
  methanol, 131
specimen handling procedures in, 131
Friel, J. J., 192, 193
Frozen hydrated image, brewers yeast, 227
Frozen hydrated samples, beam damage, 229
Frozen hydrated specimens, 224
Frozen hydrated tea leaves, chromium sputter coating, 285

G
Gamliel, H., 256
Gas plasma, earth magnet, 271
Gastro-digestive system, 245
Gatan MiniCl cathodoluminescence imaging system, 194
Gerrits, P. O., 57
Giannuzzi, L. A., 277
Giberson, R. T., 157
Glang, R., 262, 289
Glass bell jar, 267
Glass transition temperature, 285
Glauert, A. M., 48, 51, 52, 60, 154
Glues, for dry samples, 42–43
Glycol methacrylate (GMA)
  dehydration schedule using, 57
  embedding schedule for, 58
Goat anti-mouse IgG antibody, gold-conjugated, 282
Gold evaporative coating, polished mineral, 284
Gold–palladium coating, thin layers of, 273
Gold particles, secondary electron image of, 188
Goldstein, J., 2, 101, 138, 185, 188, 193, 199, 256, 274, 298
Goodhew, P. J., 79, 81
Goose Bay iron meteorite, 189
Grained igneous rock, polished sample of, 193
Green tobacco leaf mesophyll cells, 229
Grinding, 74–78

H
Hariharan, H., 13
Hawes, P., 162
Hayat, M. A., 38, 48, 301
Hayles, M. F., 90
HeLa tissue culture cells, 199
Hemicelluloses, 207
Hermann, R. J., 279, 290
High energy electrons, 2
High-vacuum evaporation, generic protocol, 266
Hirschberg, R. M., 82
Histochemical staining, 217
Hobot, J. A., 58
Hohenberg, H., 41
Holt, D. H., 49
Horobin, R. W., 57
Human astrocoma tissue, BSE image of, 283
Hunziker, E. B., 60
Hydration shell, 114
Hydrazine, 256

I
IFSM. *See* International Federation of Societies of Microscopy
Immunocytochemical analysis
  freeze substitution for, 224
  generic preparation technique, 222
  stabilization procedure, 223
Immunocytochemistry, macromolecules, 199
Immunogold-labelled cell, bacteria *E. faecalies*, 219
Impermeable specimens
  embedding and mounting of
    cold methods, 50–51
    hot methods, 49–50
Incident beam energy, 249
Industrial methods, electroplating/anodization, 259
Inert gases, nature of, 270
Infrared camera, 14
Ingram, P., 102
International Federation of Societies of Microscopy, 307
Ion beam etching, 86, 205
Ion beam guns, 85–86
  dedicated ion milling instruments, 86–87
  dual secondary electron and ion beam instruments, 89–90
  focused ion beam instrument, 87–88
Ion beam milling, 86, 203
Ion beam thinning devices, 203
Isothermal desorption, 115

J
Jacks, T. J., 106
Johansen, B. V., 290
Joy, C. S., 252
Joy, D. A., 252
Joy, D. C., 49

K
Kellenberger, E., 72
Klomparens, K. L., 301
Knoll, M., 247
Krishna, S., 277

L
Lamb, J. C., 102
Lane, W. C., 98
Lee, W. E., 49

Leipins, A., 102
Lewis, P. R., 48, 51, 60, 154
Lewis, R., 233
Lichlerberg, A. J., 86
Lieberman, M. A., 86
Liquid embedding material, 47
Liquid plastic, 41
Liquid resins, 50
Liquid suspensions, 39
Login, G. R., 156
London resins, 57
Lowicryl resin, 57, 59
  composition of, 60
  dehydration and embedding procedure for, 61
  low temperature polymerization procedure for, 61
  polymerization of, 60
LR Gold resins, 57
  embedding procedure for, 59
LR White resins, 57
  embedding procedure for, 59
  thermal polymerization of, 58
Luft, J. H., 55
Lyman, C. E., 193

M

Macromolecular analysis, immunocytochemistry, 218
Maisall, L. I., 289
Marshall, A. T., 284
Maunsbach, A. B., 301
McCann, P., 294
McGeoch, J. E. M., 90
Mechanical grinding, 74
Mechanical polishing, 75
Metal tubes, 40–41
Michael, J. R., 191
Millipore filters, 39
Morgan, A. J., 215, 224, 227, 232, 233
Muller, L. L., 106
Muller, M., 285
Murakami, T., 256
Murakami, T. H., 255
Murphy, J. A., 255, 256

N

Namork, E., 290
Natural abrasives, 76
Negative surface charging, 250
Newman, G. R., 58
Non-conducting materials, 33
Non-conductive materials coating, physical properties of, 266
Non-conductive samples coating, evaporation techniques for, 260
Non-magnetic metals, 34
Non-metallic materials, 34

O

ODO. *See* Osmium–dimethylsulfoxide–osmium
OH. *See* Osmium–hydrazine
Organic based adhesive glue, 42
Organic filters, 39
Organic polymers, 50
  synthetic, 5
Osawa, T., 281, 282
Osmium–dimethylsulfoxide–osmium, 256
Osmium–hydrazine, 256
Osmium metal coating layer, grain size, 281, 282
Osmium metal sputter coating, 280
Osmium plasma coater, 280, 282
Osmium tetroxide, 142–143
Osmium–thiocarbohydrazide–osmium (OTO) method, 255
Oxygen plasma etching, 83

P

Paints, for dry samples, 43
Paulson, G. G., 286
Peace, G. W., 290
Penning-ion beam sputter coater, diagram of, 278
Permeable specimens
  embedding and mounting of
    acrylic resins, 56–61
    epoxy resins, 51–56
Peters, K.-R., 290, 296
Petzow, G., 48
Phosphotungstic acid, 145
Pierce, R. W., 286
Planar quartz crystal, thin-film measurement for, 290
Plasma etching, 83, 238
Plasma magnetron coaters, 273
Plasma magnetron sputter coaters, 271, 290, 291, 293
Plastic film, 37
Platinum films, 265
Platinum–iridium–carbon mixture, 266
PMMA. *See* Poly(methylmethacrylate)
Polyclonal antibodies, heterogeneous, 219
Polyethylene terephthalate (PET) fiber, 67
Poly-L-arginine, 38
Poly(methylmethacrylate), 302
Polyvinylpyrrolidone (PVP), 164
Porous minerals, 47
Postek, M. T., 256
Prewett, P. D., 298
PTA. *See* Phosphotungstic acid

Q

Quamme, G. A., 232
Quantum dots
  diagrammatic representation of, 197
  elements and molecules, 196–197

Quartz thin film, 289
  practical use of, 290

R

Radioactive labeling methods, 197
Rainforth, W. M., 49
Randle, L., 191
Rat bone marrow tissue, secondary/backscattered images of, 218
Reactive ion beam etching (RIBE), 86
Reactive ion etching (RIE), 86
Richter, T., 72
Rosenberg, M., 57
Russell, L. D., 38, 43, 301
Ruthenium tetroxide, 143–144
Ryazantsev, S. N., 263

S

Sample artifacts
  faulty secondary electron images, 300
  operational pathway, 306
  organic polymers/biological samples, 301
  SE images, 303, 304
Sample cleaning
  hard dry inorganic materials, 239–242
    cutting process, 239
    hydrocarbons, 240
    microscope vacuum system, 240
    ultrasonic cleaner, 240
  hard dry organic specimens, 242
  high spatial resolution imaging, 235
  metallic samples, 237
  metals/alloys/metallic materials, 237–239
  metal surface, 239
    complicated method, 239
  plastics and polymers, 242–245
  SEM, 235
  types of
    contact cleaning, 236
    non-contact cleaning, 236
  ways of, 237–238
  wet and moist samples, 245
Sample contamination, visual indicators of, 243
Sample damage, 299
  faulty secondary electron images, 300
  organic polymers/biological samples, 301
Samples/specimens, used for SEM evaluation
  categories of
    biological organisms and materials, 5–6
    hard, dry, inorganic materials, 3–4
    hard or firm, dry natural organic materials, 4–5
    metals, alloys, and metallic materials, 3
    synthetic organic polymer materials, 5
    wet and liquid samples, 7–9
  collection of, 11–12
  dehydration procedures for, 97, 134–135
    biological organisms and materials, 136
    hard and firm, dry natural organic material, 135
    hard, dry, inorganic materials, 135
    metals, alloys, and metallic materials, 135
    synthetic organic polymer material, 135–136
    wet and liquid materials, 136
  embedding and mounting procedures for
    biological organisms and materials, 63
    hard and firm, dry natural organic material, 62
    hard, dry, inorganic materials, 62
    metals, alloys, and metallic materials, 61–62
    synthetic organic polymer material, 62
    wet and liquid materials, 63
  embedding media for, 47
  gentle mechanical and physical methods for exposure and cleaning
    gas-borne particle abrasion, 78–79
    mechanical thinning, 79–80
  high energy particles for exposure and cleaning
    combined plasma etching and ion beam etching, 92–93
    ion beam etching, 83–84
    plasma etching, 83
  methods for removal of liquids in
    air drying, 101–103
    chemical dehydration, 104–106
    critical point drying, 106–113
    freeze-drying, 114–123
    freeze substitution, 124–130
    isothermal freeze stabilization, 133–134
    low temperature dehydration, 132–133
    low temperature drying, 113–114
  methods using chemicals for exposure and cleaning
    chemical polishing, 81
    chemical thinning, 80–81
    electrochemical polishing, 81–82
    surface replicas and corrosion casts, 82
  mounting of, 47
  parameters governing selection of
    internal dimensions of SEM specimen chamber, 12–17
    large specimens, 17
    microscope operating conditions, 17
    sample shape, 18
    sample size, 17
    sample view, 18
    small specimens, 17–18
  preparation tools and associated perquisites for, 26–29
  procedure for
    labeling, 18
    storage, 29

Samples/specimens, used for SEM evaluation (*Continued*)
   rigorous mechanical and physical methods for exposure and cleaning
      by breaking, cleaving, snapping, and pulling, 66
      lapping, 74–78
      sample cutting, 71–73
      surface chipping, 70
      surface fracturing, 66–70
   sources of, 19
   suggested procedures for exposing, 94–95
   support, functions and categories of, 31
      primary supports, 32–33
      secondary supports (*see* Secondary supports, for samples)
      self-support, 32
   tools for
      cleaning, 26
      exposing, 22–23
      holding large samples, 25–26
      manipulating, 24–25
Sample stabilization, 185
   chemical analysis, preparing samples for, 230–232
   chemical composition of, 200–201
   chemical intervention prior to analysis
      histochemistry procedures, 199
      immunocytochemistry, 199–200
      quantum dots, 196–197
      radioactive labeling methods, 197
      staining, 198–199
   chemically analyze samples
      backscattered electrons, 185
      sample parameters, 187
      signals different, comparison of, 186
      x-ray photons, 185
   chemicals added, prior to analysis
      backscattered electron imaging, 187–190
      cathodoluminescence, 194–196
      EBSD, 190–191
      secondary electron imaging, 187
      x-ray spectroscopy, 191–194
   general rules, 200
      ambient-temperature wet chemical, 213–221
      environmental SEM, 212–213
      hard, dry inorganic samples, 204
      immunocytochemical methods, 205
      immunocytochemistry, 207
      ion beam milling, 203–204
      judging criteria, 202–203
      low temperature methods, 221–229
      mechanical polishing, 203
      microelectronic devices, 205
      oxide/corrosion, surface layers, 204
      sample preparation, strategy, 201–202
      staining/histochemistry, 206–207
      synthetic organic polymer materials, 208–211
      x-ray microanalysis, 206
Saws, suitable for preparing samples for SEM, 73
Sawyer, L. C., 140, 141, 210
Scanning electron microscopes
   acceleration voltage of, 190
   auto Carbon Coater, 269
   autoradiography, 197
   cathodoluminescence image, 195
      granite, polished piece of, 195
   chemical differentiation, 207
   chemical procedures for sample stabilization for imaging in
      biological organisms and materials, 146–157
      hard, dry, inorganic materials, 139
      hard, dry, natural organic materials, 139–140
      metals, alloys, and metallized specimens, 138–139
      synthetic organic polymer materials, 140–141
      wet and liquid samples, 157–158
   coating thickness for, 286–296
   depends on, 198
   dried milk powder, 276
   EBSD, 190
   electron beam, 298
   environmental, 212–213
   enzyme histochemistry, 200
   equipment to facilitate sample preparation for, 22
   Everhart-Thornley (E-T) detector, 187, 188
   focused ion beams (FIB), 276
   freeze dried, 222–223
   frozen-hydrated, 226
   Gatan MiniCl cathodoluminescence imaging system, 194
   hard inorganic materials, 240
   high resolution, 242
   high resolution field emission, 298
   high vacuum evaporation coater, 265
   image aberration/damage, 300
   imaging system, 1
   inside of chamber, 250
   ion beam sputtering, 277
   low-temperature, 213
   methods for examining wet, moist, liquid samples in, 98–101
   mica, 247
   microscopy and analysis, journals, 309
   nanotechnology, 297
   non-biological material, 195
   non-chemical procedures for sample stabilization for imaging in, 159–160
      quench cooling, 164–172
   oil diffusion pumped vacuum system, 236
   organic polymer materials, synthetic, 208
   polymer images, 209

principal approaches, 205, 208
for producing magnified images, 1
quantum dots, 196
sample preparation, sources of, 310–315
scrupulously clean, 235
SE signal, 187
signals analysis, 219
specimen preparation laboratories for
    general, 21–22
    types of, 19
specimen sizes suitable for, 20–21
thermal conductivity, 254
thermal conductivity/electrical resistance, 254
verbal information
    courses and workshops, 308
    international societies, 307
    scientific societies, 307–308
written information, 308–309
x-ray microanalysis, 194, 204, 248, 264, 284
x-ray photons, 192
Schenk, R. K., 60
Scott, R. D., 69
Secondary electron image, 199
    frozen hydrated fracture face, 228
Secondary electrons signals, 1, 137, 185
Secondary supports, for samples
    attachments with primary supports
        adhesive tapes, 43–44
        bio-organic materials, 44–45
        chemical, 42
        glues for dry samples, 42–43
        mechanical, 42
        paints for dry samples, 43
    composition and form of, 33
        fine hollow metal needles, 41
        light microscope glass slides, 38
        metal foil wrappings, 41
        metal tubes, 40–41
        mineral and plastic discs, 40
        organic and metallic filters and meshes, 38–40
        polymerized plastics, 41–42
        thin metal foils, 34
        thin supporting films for grids, 35–38
        transmission electron microscope grids, 34–35
SE images
    of artifacts, 303
    for badly charging sample, 251
    beam damage, effect of, 302
    polycarbonate grooved optical disc, 303
Sela, J., 294
SEM. *See* Scanning electron microscopes
SE signals. *See* Secondary electrons signals
Silicon, photo-resist layer, 283
Silicon substrate, high resolution images for, 241
Silver film, TEM images for, 263

Silver nitrate-photographic fixer, 256
Sims, P. A., 82
Smith, D. L., 262, 289
SNP. *See* Silver nitrate-photographic fixer
Solid embedding material, 47
Spurr's resin, 55
    infiltration schedule for, 56
Sputter coating
    argon gas inlet, 275
    artifacts and damage
        contamination, 296
        decoration artifacts, 296
        film adhesion, 296
        surface etching, 296
        thermal damage, 295–296
    ceramic matrix, 297
    final stages of, 275–276
    generic protocol, 274, 275
    ion beam sputtering, 276
        focused ion beams (FIB), 276
        turbo-molecular pumped, 276
    plasma magnetron, advantage of, 276
Stabilization protocols, precipitating reactions, 215
Staining
    heavy metal
        osmium tetroxide, 210
        phosphotungstic acid, 210
        polymer analysis, 211
        polymer functional groups, 210
        ruthenium tetroxide, 210
        uranyl acetate, 210
    histochemistry procedures, 199
    of polymers, 208
    positive/negative staining, 198
Stevie, F. A., 277
Stewart, A. D. G., 276
Stokroos, L., 273
Stöttinger, B., 82
Surface charge elimination, 247
    chemical nature of, 253
        hard dry organic materials, 257–258
        soft, porous biological material, 254–256
    EDS spectrum, 251
    electrical resistivity of, 248
    E_2 values of, 249
    general ways, 252
    image artifacts and spurious x-ray, 301
    insulators, 249
    microscope modification, 252–253
    SE image, 251
    SEM, 248
    surface coating
        analytical studies, 283–285
        choice of, 265–266
        high-vacuum evaporation methods, 264–265

Surface charge elimination (*Continued*)
  low temperature microscopy analysis, 285–286
  preparing equipment, 266–270
  SEM/X-ray microanalysis, 286–296
  sputtering, 270–283
  vacuum evaporation, 262–264
 surface conductivity, 258–262
 x-ray microanalysis, 250
Surface coating, vacuum evaporation, 262
Sylvester-Bradley, C. C., 294

T

Tanaka, K., 255, 256
Tannic acid–osmium–thiocarbohydrazide (TAOTH), 256
Taylor, A. P., 228
Tea plant leaves, location of aluminum in, 226–228
TEM. *See* Transmission electron microscopy
Thermal conduction, electrical resistivity in, 258
Thermosetting materials, 49
Thin coating layer, procedures, 262
Thin films
  formation of, 259
  making of, 36
Thin layers platinum, 273
Thomas, R. S., 233
Thornley, R. F. M., 98
Tiedemann, J., 163
Tobacco sheet materials, 304
Transmission electron microscopy, 12, 201
Transmitted electron (TE) images, 305
Turbomolecular pumped sputter coater, 272

V

Vacuum evaporation
  coaters
    diagram of, 264
    high-resolution, 279
    thin conductive coating layers, SEM, 265
  coating methods, 269
  final stages of, 268
Van der Voort, G. F., 48
Vaughn, K. C., 58
Volatile organic liquids, 103

W

Walter, P., 72, 285
Wandrol, P., 188
Wang, Z. L., 297
Warley, A., 216, 232
Water based adhesive glue, 42
Wavelength dispersive spectrometer (WDS), 192
Wepf, R., 291
Wet chemical histochemistry, 230
Wet chemical staining, 230
Wet/liquid samples
  microdroplets, 232
  selective solubilization, 232
  type of, 231–232
Wheat arabinoxylem, immunolabeling of, 221
Wilder's silver stain, 199

X

X-ray emission lines, 217
X-ray mapping, 193
X-ray microanalysis, 208
  artificially added elements, 206
  compositional information, 185
  elements, binding states of, 191
  Everhart-Thornley (E-T) detector, 188
  freeze substitution, 224
    stabilization procedures, 225
  impact of, 236
  microincineration, 232–233
  naturally occurring light elements, 206
  for producing magnified images, 1
  rock sample, 237
  SE signal, 187
  tobacco leaves, 228
  topographic information, 185
X-ray photons, 1, 185, 192, 200
X-ray spectroscopy, binding states of, 191
X-ray spectrum, 190
XRP. *See* X-ray photons

Y

Yeast samples, tea tree oil, 225

Z

Zhou, W., 297